Cool Cities

Also by Benjamin Barber

If Mayors Ruled the World: Dysfunctional Nations, Rising Cities (2013)

Consumed (2007)

Fear's Empire (2003)

The Truth of Power (2001)

A Place for Us (1998)

A Passion for Democracy: American Essays (1998)

Jihad vs. McWorld (1995)

An Aristocracy of Everyone (1992)

The Conquest of Politics (1988)

Strong Democracy (1984)

Marriage Voices (A Novel) (1981)

Liberating Feminism (1975)

The Death of Communal Liberty (1974)

Superman and Common Men (1971)

In Collaboration

The Struggle for Democracy

with Patrick Watson (1989)

The Artist and Political Vision

edited with M. McGrath (1982)

Totalitarianism in Perspective

with C. J. Friedrich and M. Curtis (1969)

COOL CITIES

URBAN SOVEREIGNTY AND THE FIX FOR GLOBAL WARMING

• • •

BENJAMIN R. BARBER

Yale
UNIVERSITY PRESS
New Haven and London

Published with assistance from the Louis Stern Memorial Fund.

Yale University Press books may be purchased in quantity for educational, business, or promotional use. For information, please e-mail sales.press@yale.edu (U.S. office) or sales@yaleup.co.uk (U.K. office).

Set in Gotham type by IDS Infotech, Ltd. Printed in the United States of America.

ISBN 978-0-300-22420-7
Library of Congress Control Number: 2016958783
A catalogue record for this book is available from the British Library.

This paper meets the requirements of ANSI/NISO Z39.48-1992 (Permanence of Paper).

10 9 8 7 6 5 4 3 2 1

To Leah Barber

I have long been conscious of the climate crisis and the political challenges it presents. But it is my wife Leah's deep moral commitment to action that has moved me from discussion and debate to action. A gifted dancer and choreographer, she has devoted recent years to taking the fight for climate action to the streets. She continues now to focus on both her art and civic action, at once an artist and a citizen. She's a cool lady with a warm heart, an outsize conscience, and a crackling talent, and it is to her I joyfully dedicate this book.

"It always seems impossible, until it's done."

—Nelson Mandela

Contents

CONTENTS

PART TWO

MAKING DEMOCRACY WORK FOR POLITICS

Acknowledgments

This book is a kind of sequel to *If Mayors Ruled the World* in that it focuses the broad thesis of urban networking and political action on the specific crisis of climate change, which—as an existential threat—outweighs every other crisis we face. It was written during an extremely challenging period during which I have worked nonstop with a small team to bring a new global governance organization, the Global Parliament of Mayors, into being. A product of *If Mayors Ruled the World,* the GPM has leaned heavily on the hard work and good will of this team and on the generous support of a number of good-willed funders.

The GPM team has brought together a number of people who worked with me on the Interdependence Movement, and who remained engaged in the quest to give cities a meaningful role in global governance. These include Jackie Davis, director of the Lincoln Center Library of Performing Arts, who not only chaired the movement's executive committee for many years but was always ready with counsel, funding support, and enthusiasm. My dear friend and fellow culture specialist and artist Eileen Haring Woods has been a longtime ally who gave us a year early in the mayors project as our executive director and then rejoined us to take the project to its realization in The Hague in September 2016. Working with us in New York and from her base in Suffolk, England, Eileen proved that cosmopolitanism is a state of mind rather than an address.

The GPM team in New York has featured a remarkable group of young professionals, including Dana Kroll, Nick Carney, Mihika Srivastava, Eric Emanuelson, and my very gifted and resourceful executive assistant and team coordinator Emilie Saccone, capable and responsible far beyond her years.

I have also had the benefit of important substantive conversations with members of the GPM advisory board, including Rob Muggah of the Igarapé Institute, John Means of McKinsey and Company, Jochen Sandig of the Berlin Radialsystem Theater, Eric Corijn, director of the Brussels Institute, and my friend and colleague Emad Tinawi. I owe Mark Watts, executive director of the C40 Climate Cities, a special debt as a longtime climate activist from his days with the City of London and a contributor to our work on the GPM and to this book.

In 2016, when I was moving from the CUNY Graduate Center to Fordham University, CUNY provost Louise Lennihan was extraordinarily hospitable. Not only did she invite me to complete my visiting fellowship in the Provost's Office, she also enabled the Graduate Center to host temporarily a Society of Patrons, something that was enormously helpful to the founding of the Global Parliament of Mayors. Her friendship and collegiality are deeply appreciated, and the role of the Graduate Center in providing a home for the writing of this book are gratefully acknowledged.

My new colleague at Fordham University Law School, Sheila Foster, who established Fordham's new Urban Consortium of which I am now a research fellow and partner, represents the kind of commitment to urban justice and cities research that makes it an honor to be her colleague. I am grateful to be working with her and with Nestor Davidson and Karen Greenberg, who helped facilitate the Fordham relationship.

A group of generous foundations, by funding the preparatory research and planning for the Global Parliament of Mayors project as well as my ongoing cities research, also enabled me also to spend time on this book. These include Bloomberg Philanthropies (special thanks to Jim Anderson), the Nomis Foundation (special thanks to Heine Thyssen and Markus Reinhard), the Levitt Foundation (special thanks to Liz Levitt), and the Kettering Foundation (special thanks to David Matthews and my dear friend, former student, and able counselor John Dedrick).

Hans Joachim Schellnhuber, the Nobel laureate who runs the Potsdam Climate Institute and advises governments in Europe and North America on climate change, played a special role in the composition of this book. As a colleague and friend who keynoted our Berlin Interdependence Celebration and Forum in 2010, he invited Leah and me to an international Laureates Conference on climate change in Hong Kong in 2015. The paper I wrote for that meeting became the seed that germinated into this book. He is thus its original inspiration, and the book stands as a tribute to his global leadership in trying to bring the world to its senses on climate change.

Finally, the many mayors I have had the fortune to engage with on the way to founding the Global Parliament of Mayors have contributed directly to the themes and concerns of this book, and indirectly to the spirit of urban leadership the book represents. They include Mayor Jozias van Aartsen of The Hague, who took the needed leap of faith when he offered to host the inaugural GPM meeting, and his colleagues in the Dutch cities of Amsterdam (where with Mayor van der Laan's hospitality we organized a key planning meeting of thirty cities in 2014), Rotterdam (where I proudly work with Mayor Aboutaleb's International Advisory Board), and Utrecht (Mayor J. van Zanen). Rob van Gijzel, the former mayor of

Eindhoven, Holland's design and technology leader, has also been an inspiration

Beyond the Netherlands, many mayors deserve special thanks, only a few of whom are named here. Among mayors who are in the leadership of the newly founded GPM and who have helped inspire my confidence in mayoral leadership are Mayor Patricia de Lille of Cape Town, Mayor Peter Kurz of Mannheim, Mayor Akel Biltaji of Amman, and Mayor Mick Cornett of Oklahoma City. Mayor Won-soon Park of Seoul hosted our preliminary planning meeting in 2014. Leoluca Orlando, the three-time mayor of Palermo and a renowned author and actor, an old friend and an urban visionary and crime fighter, has been a model urban leader and terrifically helpful in planning and bringing colleagues to the GPM. Mayor Anne Hidalgo of Paris is a new friend and new mayor who in a remarkably short period has assumed a global leadership role in urban cooperation and climate change. Former mayor George Ferguson of Bristol and his successor Marvin Rees, Mayor Eduardo Paes of Rio de Janeiro, Mayor Hanna Gronkiewicz-Waltz of Warsaw, Mayor Giorgos Kaminis of Athens, and many more have given palpable form to an idea for turning cool cities into warm advocates of global cooperation. In the United States, the longtime CEO of the U.S. Conference of Mayors, Tom Cochran, has been a generous collaborator of the GPM and an avid advocate of cities. Jerry Abramson and Elias Alcantara in the White House Office of Intergovernmental Relations have been encouraging supporters of my cities work. Mayor Eric Garcetti of Los Angeles was an early reader and advocate of *If Mayors Ruled the World,* sharing it with colleagues and in effect putting it on the desks of mayors from Boston to Seattle, New York to Atlanta, Chicago to New Orleans.

Throughout the writing of this book I have had the collaboration of a dedicated and able research associate, Matt McEnerney. A student of urban affairs and now a professional consultant, Matt has worked diligently with me in surveying the environmental landscape and has contributed important empirical research, data, and bibliography. He has also been a useful interlocutor in thinking about cities and climate change. While working with me on the book, he assisted the team working on the Global Parliament of Mayors project, and he has my gratitude for these many contributions.

This is my second book with Yale University Press, to which I came through my longtime friend and colleague Steve Wasserman, as my literary agent, then a Yale acquisitions editor, now the publisher and executive director of Heyday Books back in his beloved California. I first knew and worked with Steve as the editor of my *Jihad vs. McWorld,* after he read my cover essay in the March 1994 *Atlantic Monthly* and, fatefully, persuaded me to put into book form. (My next project, "Isis on the Internet," is the sequel to *Jihad vs. McWorld.*) Once at Yale, I had the good fortune of working with Bill Frucht, an editor who quietly knows how to bring out the best in a manuscript. First *If Mayors Ruled the World* and now *Cool Cities* have benefited from his unobtrusive but careful and firm editorial scrutiny—he would have insisted I choose between "careful" and "firm" and avoid using the slightly redundant "unobtrusive" altogether, but I did not let him edit this paragraph. I must of course take full responsibility for any flaws or errors that remain in the manuscript.

INTRODUCTION

Politics Not Science

Climate change is the most urgent challenge facing humankind in the new era of interdependence. Other issues make headlines: terrorism kills, and kills now; inequality affects everyday life for billions around the globe; 60 million people have been displaced by war, oppression, and genocide, as well as a desperate need for work, and they are everywhere on the march across borders. At the same time, proto-fascist political movements threaten citizens from a disintegrating Europe and a fragmented Latin America to a polarized United States, and governance seems nearly paralyzed.

But climate is paramount, because in sustainability human survival itself is at stake. Why then have the nations governing the planet been so hopelessly ineffective in addressing the grave environmental crisis?

Is it because the consequences of carbon emissions leading to increases in atmospheric and ocean temperatures seem hypothetical, or too far off? A distant Noah's flood, but as far in the future as the ark was in the past? Politicians pay few costs for doing nothing, and receive little credit for acting aggressively. In the United States, a nation that contributes one-fifth of all global greenhouse gas emissions (China is responsible for another 20 percent), one of the two major political parties continued to deny there was even a problem during the 2016 presidential campaign. Its candidate, Donald Trump, promised to reopen coal mines and free up oil drilling. He is now the president of the United States. What some dismissed as polemical hyperbole during the campaign is now part of a prospective White House climate policy. Trump tweeted while campaigning in 2016: "This very expensive GLOBAL WARMING bullshit has got to stop. Our planet is freezing, record low temps."[1] Although he denied it in a 2016 debate with Hillary Clinton, years earlier Trump had claimed in a tweet (posted on November 6, 2012) that the "concept of global warming was created by and for the Chinese in order to make U.S. manufacturing non-competitive." Climate change otherwise went wholly unmentioned at the Republican National Convention as well as in the campaign, except for a popular sign reading "Trump Digs Coal!" Moderators failed to put a single question on climate change to the candidates in the three fall debates, and it is fair to say that the most urgent problem facing the planet was entirely ignored by the press no less than by the candidates.

In December 2015 in Paris, after twenty years of futility, the United Nations climate change conference, known as the Conference of the Parties (COP 21), finally reached a soft agreement to (try to) limit the global rise in temperature to less than two degrees Celsius. Welcome as it is, this agreement, signed on Earth Day 2016 by 174 nations, does far less than is demanded by climate scientists, who have greeted the accord with muted skepticism. Moreover, candidate Trump promised during the campaign to withdraw from the Paris agreement. Whether President Trump will make good on this threat remains to be seen, and his power to do so is not altogether clear. But the treaty is too dependent on self-reporting and too weak on enforcement to survive an antagonistic president even if he abstains from formal action.

In any case, even before the convening nations concluded their business in Paris, German climate expert Oliver Geden wrote: "As delegates meet in Paris . . . we are already in danger of busting the [carbon] budget. If the plans submitted by more than 180 governments are implemented, humanity will outspend its carbon budget by 2040 at the latest. Staying within the original budget outlined by the Intergovernmental Panel on Climate Change (I.P.C.C.) no longer seems realistic."[2]

Better news in the fall of 2016, when an amendment to the landmark 1987 Montreal Protocol banning chlorofluorocarbons (CFCs) aimed at closing the ozone hole was adopted by 170 countries in Kigali. The Kigali amendment cuts the production and use of hydrofluorocarbons (HFCs) used in air conditioners and refrigerators and, by doing it as an amendment to a treaty already in force, allowed President Obama to bypass Congress. HFCs are only a small part of greenhouse gas emissions but they have a heat absorbing capacity one thousand times that of carbon dioxide, and their elimination thus has a profound impact on warming. Unlike the Paris agreement, which is "voluntary, often vague and dependent on the political will of future world leaders . . . the Kigali deal includes specific targets and timetables to replace HFCs with more planet-friendly alternatives, trade sanctions to punish scofflaws and an agreement by rich countries to help finance the transitions of poor countries to the costlier replacement products."[3]

Yet though the Kigali amendment was welcome, the IPCC was not impressed by the Paris agreement, estimating that the two-degree limit, even if realized, will still allow a three-foot rise in sea level by the end of the twenty-first century. And the IPCC estimate is pretty conservative. The Army Corps of Engineers projects four feet, while the National Oceanic and Atmospheric Administration predicts it could be six and a half feet.[4] Other estimates range from ten to thirty feet of sea rise. Even with HFC regulation ameliorating temperature rise, as little as three feet of ocean rise will still put 200 million people who live within six feet of high tide at risk and leave scores of world cities including Miami, Shanghai, New York, London, Sydney, Hong Kong, and New Orleans underwater. Why on earth can't responsible governments take responsible and decisive action?

The problem isn't the science. The facts are indisputable. The empirically grounded analyses predicting cataclysm have grown quite stale with repetition. We have been warned again and again and again. Rachel Carson warned of spring gone silent (the price of DDT) in the 1960s; in the '70s, sober scientists at MIT invoked "limits to growth" that if exceeded would endanger the planet; then in the '80s there was the young Bill McKibben predicting the end of nature. Right into our new millennium, knowledgeable prophets have become doomsayers, the people who know the subject best, raising their voices to keep pace with rising greenhouse gas emissions, rising temperatures, and rising sea levels. Some are scientists, some journalists; some are statesmen, many more are simply citizens. They include Oscar-nominated filmmakers like John Fox (*Gasland* and *How to Let Go of the World*) and Nobel Prize–winning physicists like Hans Joachim Schellnhuber, director of the Potsdam Climate Institute in Germany. Their names are familiar, if not always as recognizable as Carson's and McKibben's: Maude Barlow, Gus Speth, Jeffrey Sachs, Naomi Klein, Michael Mann, Elizabeth Kolbert, James Hansen, and of course the champion of inconvenient truth, former vice president Al Gore. The IPCC has been issuing periodic assessments of the problem since 1990—as of 2016 they had done five—and with each new assessment the problem is more dire, warming having advanced faster than predicted in the previous report. And now this raucous earthly choir is joined by the beatific voice of Pope Francis, warning in his encyclical *Laudato Si': On Care for Our Common Home* that climate change and justice go hand in hand and that the stewards of God's bounty have misread a bounteous earth as an invitation to excess.

No further warnings are needed. Not when the earth itself manifests on a daily basis the truth of the dire prophecies; not when our follies are, as McKibben wrote, patently "altering every inch and every hour of the globe."[5] Long ago, in ancient India, an Ashoka mandate promulgated on the Indian subcontinent taught that "forest must not be burned in order to kill living things or without any good reasons." A thousand years later, we are ripping the carbon fossils of burned forests from the ground to power an insatiable hunger for wealth and dominion: "Big Oil Wants to Burn It All," inveighs a headline in *The Nation*.[6]

It is as if we have chosen to raise the dead, animals and vegetable alike, and burn them all over again, allowing the carbon and methane released to settle into the oceans or waft into a warming atmosphere. Disinterred, the molecules afflict us from above and below, a kind of molecular sacrilege destroying the planet's ecological balance. Exhuming our carbon ancestry faster than the planet can recycle it, we put the living at grievous risk.

This may seem rather grand, not carbon dioxide but metaphor unleashed, but it is only science dressed up for dramatic effect. Draped in metaphors or presented as hard data, the facts are clear and indisputable. So no, science is clearly not the problem. The problem is the politics. If we hope to do more than surrender (which requires of us only a sense of humor and a talent for despair), we must act to move beyond the dismal end-time scenarios.

The time for Panglossian techno-zealots shilling the "next shale revolution" and "the astonishing promise of enhanced oil recovery" is, like nature itself, over.[7] The merchants of doubt who once claimed that links between tobacco and cancer were "unproven" and today hint at a climate science open to scientific debate are not *scientific* adversaries at all. They are political adversaries, mostly bought and paid for.[8] Those who still insist every pernicious byproduct of progress gone wrong can be overcome by another round of progress made right have political axes to grind or profits to pursue. Listen to economist William D. Nordhaus snicker about how "little economic impact upon advanced industrial nations" climate change will have other than that "snow-skiing will be hurt . . ." But not to worry, "water-skiing will benefit."

To be sure, full-out climate change deniers do still exist. But their dogmatism persuades a minority of citizens only because of a deep cultural ignorance about the nature of science, and hence the science of nature. We live in an American civilization where a proudly ill-informed subculture insists evolution and creationism are two competing instances of subjective opinion and that "proof" is whatever you happen to believe. Yet to the extent that they are not simply fools, the deniers are creatures of convenience. They are expansive ideologues (government does nothing right! markets can do no wrong!) or shrunken capitalists looking for profits. So yes, we still hear deniers making small and cheerful denier noises, like the oblivious chirping of chipmunks before the hawk snatches them up. The hawk, however, cares neither whether his meal notices him nor whether it thinks it is in a "negotiation" with him. Nature doesn't negotiate.

What is of concern is no longer nature and its laws, but nature as occupied and shaped by humankind. As McKibben reminds us, "If the waves crash up against the beach, eroding dunes and destroying homes, it is not the awesome power of Mother Nature, it is the awesome power of Mother Nature as altered by the awesome power of man."[9] That awesome power is certainly evident in 2016, the hottest year ever recorded on the planet, in a decade of record hot years, in which Patricia, the largest hurricane ever experienced rolled across Mexico while superstorm Matthew roared up the Atlantic coast in

October. We need no longer be diverted by foolish evasions or earnestly stupid arguments about what the facts really are. Instead we must move from nature to humankind, from the firm ground of the empirical to the slippery slopes of the political. It is here that the struggle for sustainability must be won. Despite the indisputable arguments of science, the American presidency has been won by a climate denier who has threatened to transform the most important national advocate of sustainability into a deeply consequential opponent. The science is clear but the politics is decisive.

That makes this a book about politics. A book also, its dark beginning notwithstanding, about hope. The realm of politics invites cynicism, but it always engenders hope. Politics is the domain of human will, human interest, human power, and human action. It is where we make collective decisions to deal with the public consequences of our private actions. It is how we contend with the devastation of Hurricanes Katrina and Patricia; how we mobilize a half million New Yorkers to march in the streets to teach anxious pols that it is time to connect the heating that allows the atmosphere to suck up more and more water with the huge Category 5 storms it spins off in what we now call extreme weather, which is a political rather than a meteorological issue.

The Trump victory is a victory of politics. It will be overcome only through politics. This means a politics of empowered cities exercising the right to secure human sustainability, especially when nations fail to do so. Politics at its best allows us to decide together as citizens how to undo the inadvertent common effects of all that we do one by one as consumers and producers, or what we do when private interests and prejudice seize the institutions of the state to try to undo public goods. It is the arena where public goods can trump private interests, where the commonwealth can become a measure of higher purposes and the mirror of public values. But politics is hardly at its best right now, and that is perhaps the greatest challenge facing us as we confront climate change. The weakness of politics undermines democracy—the faith behind politics. Too many citizens view the political success of nativists and xenophobes like Marine Le Pen or Geert Wilders and now Donald Trump as the defeat of politics, of democracy itself.

I cannot accept that view. Since this is a book about politics, it must also be a book about democratic faith. Democracy is at once the most serious obstacle those who would address climate change must overcome, and their indispensable vehicle for achieving success.[10] Democracy corrupted turns mass opinion toward narcissism and self-interest. Democracy actualized and legitimate offers a politics of hope against corrupted democracy's politics of fear. The dialectic between the two yields this book's most troubling paradox: how can a cynical domain of power and self-interest become a domain of sustainability and justice? If money and the short-term interests of "me" drive politics, how can it secure the long-term interests of "we"? What will it take for our institutions to embody a challenging politics of hope and global purpose over the easy politics of fear and egoism? At which level of government can citizens best leverage political institutions to secure their sustainability? Democracy is crucial because climate change is also about justice: how to distribute the costs of decarbonization and a costly transition to renewable energy equally and fairly among rich and poor, developed and developing, large and small, north and south. We cannot elude the subject of justice because the costs and benefits of addressing climate change are inevitably skewed around wealth and power. The rich man reacts to the rising tide by moving his summer home from Cannes to St. Moritz. The poor woman holding her newborn drowns.

Justice is not only about social and economic relations, it is about generational and interspecies relations. The global north won its modernization and abundance through development that was unsustainable for the planet, but whose price it neither calculated nor paid. The developing world—mostly the global south—is now expected to pony up. Developmental latecomers like Nigeria or Brazil, told they cannot ground their development in carbon because the developed nations used up the earth's carbon quota on their own growth, are likely to be underwater economically for a long time; mini-nations like the Maldives, meanwhile, will be literally underwater by century's end. If fairness is to be achieved, justice will have to spread itself across continents and the generations. The new strivers who seek prosperity today cannot be asked to pay tomorrow for the costs of the old strivers who secured their bounty yesterday.

Species fairness is also at stake. How to strike a balance between humankind and our fellow species, whose needs often do not register at all on the sustainability index? Elizabeth Kolbert has estimated that climate change puts at least a quarter of extant animal species at risk of extinction in the near term. Long term, as many as half could vanish. The catastrophe of global warming is also a catastrophe for biodiversity, which is hence not a chapter in some other book, but a chapter in the book of climate change. The end of nature Bill McKibben foresaw decades ago goes by a geological name: the Anthropocene era.

The term *Anthropocene* describes a new geological era that began with the industrial revolution, whose fossil record future geologists will read as deeply marked by human activity.[11] Politically, the Anthropocene has a more tendentious edge. It calls up the peculiar combination of arrogance and obliviousness that define the dominant political paradigm—a cynical politics that responds more readily to dogmatism's illusions than to nature's realities; and that makes common cause, if at all, only in the name of private interests that deny the claims of commonality.

The antidote is to be found in a politics of hope and purpose. But, as I argued in *If Mayors Ruled the World,* that politics cannot be found in (or rescued from) increasingly dysfunctional nation-states or rigidly ideological national political parties. Locked inside a sovereignty defined by an independence that feels today more parochial than cosmopolitan, national leaders have defaulted in dealing with the big issues of an interdependent world: terrorism, anarchic financial markets, pandemic disease, nuclear proliferation, the refugee crisis, economic justice, and climate change.

National states and their fractious political parties are undermined in their democratic aspirations not only by parochialism but by money, media, and manipulation. They boast that they represent the citizenry, but many citizens regard them as bogus. They run national political campaigns in the name of an establishment that voters disdain. But in rejecting the posturing "democrats" whom they see as oligarchs barely disguised, voters flee toward the seductive demagoguery of populist pretenders aspiring to take over that very establishment.

The alternative to a politics of cynicism is a politics of participation that devolves power back to people closer to where they actually live: back to cities. Shift the focus down to municipalities and over to civil society, and recognize that mayors are in an ideal position to understand the real problems people have, and to implement real solutions. Give them the authority and resources to craft a sustainable future.

If sustainability demands decarbonization, decarbonization demands devolution. Hope for the future lies with the politics of the city. From the pragmatism of mayors and their capacity for global cooperation, which nations have lost (and perhaps never had) and cities have found (and perhaps always had), we can generate a sustainable politics for a warming world. I made the case for cities as global players in *If Mayors Ruled the World.* Here I hope to show their political aptitude is compellingly manifested in their potential as agents of sustainability and resilience. On a hot planet, cities are cool.

I am hardly the first to notice this, or to argue that urban politics can make the critical difference in combating climate change. Mayors are ardent urban boosters. Former San Antonio mayor Julian Castro (President Obama's housing secretary) says, "Cities are where the future happens first."[12] New York's former mayor Michael Bloomberg has insisted for some time that "cities have played a more important role in shaping the world than empires."[13] U.N. secretary general Ban Ki-moon, whom Bloomberg now serves as climate emissary, adds: "Cities can be the engine of social equity and economic opportunity. They can help us reduce our carbon footprint and protect the global environment." Pope Francis carries the logic beyond its moral to its political conclusion. In a prudent political voice, he appeals to the "urgent need" for "a true world political authority."[14] This book is about how we institutionalize that "true political authority" as a secular means to Francis's moral vision of a just environmental sustainability.

Even with Donald Trump in the White House and the tepid but important Paris climate agreement in jeopardy, a truly democratic politics of cities that are knit together in global networks of collaboration holds out a promise of environmental redemption. Collaborative urban politics encompasses networks of every kind, from the United States Conference of Mayors and EuroCities to United Cities and Local Governments (UCLG) and the Global Parliament of Mayors outlined below. It promises that what we do ineptly and perversely in the dysfunctional setting of national politics and consumer narcissism today, we can undo and overcome tomorrow through common civic action in a networked urban cosmopolis. Having made the Anthropocene, we can unmake it. Having ended nature (or more accurately, put it on "pause"), we can right the balance between its requirements and our own. It is as citizens of functioning cities that we may set agency against destiny and politics against necessity. Using the power we have borrowed from nature to master it, we can restore harmony and justice to what can once again become a relationship of reciprocity. As we devolve authority and confront a shift in the locus and impact of sovereignty on this new interdependent planet, as we look from dysfunctional nations to rising cities, we have an opportunity to renew the social contract and rewrite its terms.

PART ONE
Making Politics Work for Science

• • •

"2015 offers a great opportunity. Let it be the year that cities clearly assume the mantle of leadership in getting the world on a sustainable path."

—Eduardo Paes, mayor of Rio de Janeiro and former chairman, C40 Cities Climate Leadership Group

1

The Social Contract and the Rights of Cities

The science of human survival is political science. Survival depends on sustainability and resilience, and the means to sustainability and resilience are political. It is for survival (security) that naturally free human beings enter into a social contract and bind themselves to obey the sovereign governing bodies they establish. Centuries ago, when the idea of a social contract was established in the West, the sovereign governing bodies able to secure life and liberty were conceived as nation-states. But as the world has become more global and interdependent, sovereign nations and their international networks have grown less effective, sometimes even dysfunctional. Survival—a sustainable world—depends more and more on citizens acting locally in the name of global goods, of which climate change and decarbonization are prime examples. Sustainability today entails *glocality,* action that is simultaneously local and global. Municipal policies must be crafted with an eye on their impact not over months or even years but over generations, as well as among communities and peoples across the interdependent planet.

Of the many threats to a sustainable world, none is more dramatic and perilous than human-induced climate change and its consequences, which include global warming, sea-level rise, and extreme weather. The collective impact of these consequences is putting civilization at risk—indeed, perhaps putting life on earth at risk. For even though as Lynn Margulis liked to say, "Gaia is a tough bitch," whether the planet is tough enough to deal with our species' hubris is yet to be seen. I propose in this volume to address climate change by focusing on municipal approaches to renewable energy and a non-carbon economy, to decarbonization in a metropolitan setting. Cities can do decarbonization, and when they act interdependently, they can do it on a scale relevant to global warming.

The agency and actions needed are urban and local rather than national. Cities are home to more than half of the human population and more than three-quarters of the population of developed nations. They generate 80 percent of global GDP as well as 80 percent of greenhouse gas emissions. They also suffer the lion's share of the economic damage from extreme weather events and sea-level rise. Along with agriculture, they consume much of the planet's water, and the metropolitan regions they define house the factories and plants that run on carbon energy and account for a preponderance of carbon emissions. Private-sector automobiles and trucks are massively polluting, and public transit systems, unless they are upgraded and electrified, make things worse. The density and lack of green space in cities make them an environmental problem from the get go. Yet density also gives them a smaller collective carbon footprint per capita than suburbs or rural regions. Cities are the problem. But cities, as both the prime sources and prime victims of climate change, can also be agents of remediation: politics at the municipal level may prove the equal of climate change at the global level. We can take the solution into our own hands. Whether we will is the question of the hour, and of the millennium.

We certainly are not likely to succeed if we depend on aging nations, which have already proved their deep incapacity to address global problems. Their dysfunction runs deep, and democracy appears to be in retreat, not only in the obvious places like Russia but also in Brazil, where an entire government was impeached in 2016 just before the Olympics; in the Philippines, where a demagogic mayor has become president; in Austria, Hungary, Poland, and Belarus, where populist and antidemocratic governments have been democratically chosen; and in the United States, where partisan gridlock has paralyzed governance, while a divisive and angry presidential election in 2016 left many American citizens disturbed and fearful for the democratic process itself. The crisis in national governance is a crisis in sovereignty, in the capacity of the nation-state to make good on the terms of the social contract on which their founding legitimacy turns.

Ideally, good-willed national states and international governing bodies would acknowledge their growing sovereign inadequacies and welcome, rather than condemn, urban efforts to solve global problems whose remedy has eluded them. In practice, that has not always been the case.

In federal states with a vertical separation of powers, such as India, Germany, Brazil, Canada, and the United States, local governments more readily accept power sharing. Although the classical theory of sovereignty leads unitary governments with hierarchical power structures to be chary of local autonomy, even strong top-down administrations sometimes tolerate a certain municipal outspokenness. Nonetheless, they hold strongly to the view that cities must finally be regarded as administrative subsidiaries of the state. In France, England, and China, for example, cities enjoy little jurisdictional autonomy. Deprived of both resources and jurisdictional competence, they are unable to walk an independent path.

This stern view of the subsidiarity of cities has softened recently as a result of an emerging recognition that cities do get things done and that the devolution of authority can benefit national as well as local government. The relationship between city and state cannot, however, be left for nation-states alone to define. As the representatives of citizens, who are in turn the ultimate source of sovereignty, municipalities understand that their claim to jurisdiction rests not only on their *capacity* to act effectively but on their *right* to do so when higher jurisdictions fail to discharge the responsibilities of sovereignty. This conundrum returns us to the foundational idea of the social contract.

The right of cities to govern themselves and to come together with other cities, both within and beyond their national borders, increasingly is being grounded in powerful rights arguments, even though many local government officials are loath to appeal to them too boldly. Many prefer to justify cross-border collaboration without making rights claims at all. There are many reasons for cities to join urban networks that promote common urban policies, and even to risk joining an explicit governance body such as the Global Parliament of Mayors, without appealing to principles of right. But when cities face serious resistance to their acting in common and across national borders, the rights argument offers significant support.

For this reason, I ask readers to indulge a brief detour through social contract theory. The right of a sovereign to govern derives from the right of the individuals who contract with one another (in the "state of nature") to establish that sovereign to protect them. It is important to note here that the term *sovereignty* has a precise meaning. It is defined formally in classical political theory (Hugo Grotius and Thomas Hobbes, for example) as the rightful exercise of power by a governing body whose legitimacy depends on its capacity to secure the life, liberty, and property of its citizens. This is why we accept the sovereign's laws, pay its taxes, risk death or injury serving in its military, suffer the punishments meted out by its courts, and honor its leaders: in exchange, the sovereign must guarantee the sustainability of both society and its individual citizens. It must protect us from foreign aggressors, domestic criminals, and natural and environmental threats. Citizens obligate themselves to obey the sovereign because the sovereign obligates itself to secure their lives, liberties, and properties. So say Locke and Rousseau; so reads the Declaration of Independence.

But by this logic, the authority of the sovereign lasts only as long as it keeps citizens safe. When cities act together across borders in the name of securing the lives of citizens put at risk by the failure of states, they are in effect asserting there has been a sovereign default by states and reclaiming their sacred right to act as the surrogate sovereign on behalf of their citizens. By failing to protect us from climate change, nation-states have reneged on their end of the bargain. Their inability to assure sustainability constitutes a default of sovereignty that both permits and demands action from alternative authoritative bodies: in our situation today, from regional and municipal authorities who assume responsibility for sustainability. If states can't or won't assure sustainability, cities must. It was just this logic that led American colonists living under the sovereignty of an English king to make clear the basis of their allegiance. In language that resonates today, the colonists declared:

> We hold these truths to be self-evident, that all men are created equal, that they are endowed by their Creator with certain unalienable Rights, that among these are Life, Liberty and the pursuit of Happiness.—That to secure these rights, Governments are instituted among Men, deriving their just powers from the consent of the governed.

And then, in the famed revolutionary sequitur, they concluded that their obligation to obey was circumscribed:

> Whenever any Form of Government becomes destructive of these ends, it is the Right of the People to alter or to abolish it, and to institute new Government, laying its foundation on such principles and organizing its powers in such form, as to them shall seem most likely to effect their Safety and Happiness.

It would go too far to suggest that the failures of modern nation-states have robbed those states of their sovereign right to govern. The United Nations is hardly George III. Yet it is precisely the conundrum of modern nation-states and the international bodies they establish that they can no longer assure sustainability in the face of rising greenhouse gas emissions, sea rise, and extreme weather. They are undermined by a fatal asymmetry between their sovereign borders and the borderless nature of the climate challenge they confront.

Cities, too, partake of the social contract, if only tacitly. They tax and police their citizens, pass laws, and punish lawbreakers. This imposes on them a parallel obligation to that of states: they must protect their citizens' life, liberty, and property and endeavor to ensure sustainability. And an obligation implies the right to discharge it. Undergirding the obligation of cities to deploy common power on behalf of sustainability, then, is an implicit right to do so on behalf of their citizens. Think of the consequences of climate change unmet, or a global refugee crisis (60 million people in motion) unresolved; think of cross-border terrorism aimed almost exclusively at cities with no remedy in view; think of nuclear proliferation as an unstoppable tide; think of global markets, pandemic disease, and the peril brought by extreme social and economic inequality within and among nations if they are left without responses.

Such questions are anything but hypothetical in the era defined by Brexit and Donald Trump. In anticipating abusive action by a

Trump administration, Mayor Bill de Blasio of New York was astonishingly assertive about the rights of the city and its will to resist. In blunt comments made at Cooper Union two weeks after the election, Mayor de Blasio said:

> We will fight anything we see as undermining our values. And here is my promise to you as your mayor—we will use all the tools at our disposal to stand up for our people: If all Muslims are required to register we will take legal action to block it. If the federal government wants our police officers to tear immigrant families apart, we will refuse to do it. . . . If the Justice Department orders local police to resume stop-and-frisk, we will not comply. We won't trade in neighborhood policing for racial profiling. If there are threats to federal funding for Planned Parenthood of New York City, we will ensure women receive the healthcare they need. If Jews, or Muslims, or members of the LGBT community, or any community are victimized and attacked, we will find their attackers, we will arrest them, we will prosecute them. This is New York. Nothing about who we are changed on Election Day.

These are literally fighting words, invoking an implicit local sovereignty and right of resistance. Moreover, de Blasio makes explicit the constitutional argument on which his defiance is predicated.

> There is not a national police force. You don't go to federal schools to get your children an education. . . . We in the City of New York, we protect our people . . . In the Declaration of Independence there is one of the most simple and powerful passages: "governments are instituted deriving their

just powers from the consent of the governed." We don't consent to hatred.

The only thing missing is the necessity for intercity cooperation and networking. Cities cannot succeed one by one. Interdependence requires a common urban voice.

Urban networks operate by law in the shadow of national sovereignty, which mandates that cities remain the political subsidiaries of sovereign national governments. Yet these governments face the specter of sovereign default. This default legitimizes cities in convening across borders and claiming the right (and hence the jurisdiction and resources) to sustain the lives and liberties of their citizens. Challenges without frontiers demand civic action without frontiers, political bodies without frontiers—the very definition of modern cities on an interdependent planet.

Behind the convening of a Global Parliament of Mayors stands this foundation in rights. Although subsidiary in law to the sovereign nations to which they belong, cities do not require "permission" to act collaboratively—although they may well conclude it is prudent to ask for it. The Hague Declaration that accompanied the founding of the Global Parliament of Mayors in 2016 affirmed that the new organization "will, wherever possible, cooperate with nation-states and international networks and organizations, especially the United Nations."[1] At the same time, the GPM Mission Statement introduces the new body as a "cities rights movement." Sovereign default, a consequence of interdependence, is leaving nation-states with diminished democratic legitimacy while endowing cities with ever increasing democratic authority. Behind that authority sits the real sovereign—the citizen—who in democratic republics is always the source of political authority at whatever level it is exercised. Social contract theory holds that although power may be exercised top-down, its legitimacy flows from the bottom up. In the new study *City Power*, Richard Schragger suggests that even if they do not currently govern, "we should *want* cities to govern," and that the "*desirability* of city power" brings with it a claim to the power and resources required for them to govern.[2]

It is not just a matter of the desirability of city power. They already exercise default jurisdiction in many domains arising out of a default in nation-state. Moreover, compelling as the sovereign default is, it is reinforced by a long history of charters that include the rights of cities and justify the rights of urban citizens to take matters into their own hands. This history suggests that the Global Parliament of Mayors is only the latest of many municipal networks claiming the right to associate and act across borders, going back to the ancient Mediterranean League of Cities and the medieval Hansa, or Hanseatic League of cities along the North Sea, and continuing through the century-old network United Cities and Local Governments, the C40 Cities Climate Leadership Group, and scores of others. At the same time, the GPM is not just another discretionary intercity association but a milestone in the history of networks, an effort to effect a true governance revolution that establishes a novel form of political authority, rooted not only in capacity and responsibility but in the universal right of citizens to guarantee the sustainability of their own lives.

The key rationale for the Global Parliament of Mayors (GPM) is the right of self-governance. This founding right has been historically recognized in documents as old as Magna Carta, which had its eight hundredth anniversary in 2015 and whose ninth paragraph reads:

> The city of London is to have all its ancient liberties and customs. Moreover we wish and grant that all other cities and

> boroughs and vills [*sic*] and the barons of the Cinque Ports and all ports are to have all their liberties and free customs. . . .

The American constitution clearly states that "the Laws of the United States . . . shall be the supreme Law of the Land," but the Bill of Rights underscores both limited government and the federalist separation of powers. It references the same local rights upheld in Magna Carta, both in the Ninth Amendment, which recognizes the people as the ultimate repository of rights, and in the Tenth, which reserves those powers not explicitly delegated to the federal government to the states and the people.[3]

Local authority is also recognized in the European Charter; its version of the doctrine of subsidiarity places states above cities but endows cities with key governance rights not dependent on the central government. This history of municipal and citizen rights lies behind Alexis de Tocqueville's insistence in *Democracy in America* that liberty is local, and his celebration in *The Ancien Regime and the Revolution* of the rights of French provincial *parlements.*

The right of the city and its citizens to self-government is also legitimated by the principle of majoritarian democracy: the global majority today is urban. It finds further support in the principles of limited government and of federalism and confederalism, which call for power to be shared not just horizontally (through the separation of powers) but vertically, and which treat local and municipal government as coequal civic domains. The common etymology of the words *city* and *citizen* (*cite* and *citoyen*), as well as *Burg* and *Bürger, polis* and *politics,* shows yet again the intimate link between the city and civic self-governance.

Add to democratic legitimacy the economic legitimacy that attends another well-known urban fact: as I mentioned above, more than 80 percent of global GDP is generated in cities. In the United States, it is 90 percent of GDP, as well as 86 percent of jobs.[4] Even greater proportions of science, culture, patents, and education are concentrated in municipalities, which is why our cities have always been civilizational nodes. I am not suggesting that this makes cities "better" than other habitations or gives them the right to govern at rural areas' expense. But those who make up a majority of the population and generate most of its wealth, cultural products, philanthropy, education, and inventions have a right to some degree of self-governance and to political and civic institutions that assure their sustainability and survival.

The size of urban economies often surprises casual observers: New York City has a larger economy than Spain, South Korea, or Mexico; Philadelphia's economy is bigger than those of Venezuela and South Africa; Miami–Fort Lauderdale (Florida) is in front of Chile, Finland, and Egypt; Charlotte, North Carolina, outranks Hungary; Tulsa's numbers are larger than Ghana's, and Albuquerque is in front of Turkmenistan.[5] The urban tail often wags the national dog: Cairo's GDP, at over $100 billion, is nearly one-third of Egypt's ($331 billion); while Bangkok's $119 billion is more than a quarter of Thailand's ($400 billion).[6] And New York is only the second most productive city in the world. According to Richard Florida's Global Economic Power Index, Tokyo is number one, with $1.2 trillion in economic output, while London is number three (just after New York), followed by Chicago, Paris, Boston, Hong Kong, Osaka, and Seoul at number 10. Other American cities come in at four (Chicago) and number six (Boston).

None of these arguments guarantee that nations will passively stand by while cities assert rights claims, however legitimate. Nor should cities brashly tell nations off. Cooperation and complementarity are not only politically prudent but are best suited to the nuanced transactional character of cities and their appropriate disregard for force (they have neither armies nor navies—although Mayor Bloomberg once boasted the NYPD *was* a kind of army). But cities should move ahead confidently when cooperation is spurned and their cross-border initiatives elicit truculence and resistance from nations, or litigation aimed at crippling their actions, as happened in the United States with court decisions preventing cities from protecting citizens by banning fracking or registering guns. Citizens are fighting back in the courts.

Individuals in Pakistan and New Zealand are suing their governments to take action on climate change, while in the United States some courts are acknowledging the right of citizens to protection against climate change. In the spring of 2016, Thomas M. Coffin, a federal magistrate judge in Oregon, "startled many legal experts by allowing the lawsuit filed on behalf of 21 teenagers and children to go forward, despite motions from the Obama administration and fossil fuel companies to dismiss it," according to a report in the *New York Times*. Michael B. Gerrard at Columbia Law School, who succeeded Jeffrey Sachs as director of the Earth Institute in 2015, commented that the ruling "suggested that government may have a constitutional duty to combat climate change, and that individuals can sue to enforce that right."[7] The notion that citizens and hence their cities have a right to protection against climate change is as radical and consequential as it seems—and if courts begin to support the argument, it is potentially transformative.

Acting against the command of a sovereign state, however porous the sovereignty, should always be a last resort for cities, and then only when they have been blocked at every turn in securing their sustainability. At that point, they can do what individuals and popular movements have always done when their rights are denied by arbitrary authority or blocked in the name of sectarian and private interests disguised as national goods: they can act. To be successful, they will have to do what rights-seekers always do: appeal first to courts, as happened in Oregon, and then ground their legal struggles in urban party movements, popular resistance, and direct citizen actions, and where other strategies fail, in civil disobedience.

If a city bans guns (as Washington, D.C., did in 2013) and watches a higher jurisdiction's court strike down its statute, it should persist in its actions and invite the higher jurisdiction to sue (although Washington has more fundamental rights issues, lacking standing as a state or a city). If a state or national government nullifies a city's decision to ban fracking or block a gas line, that city can take to the courts and the picket lines—ideally in concert with other cities. A national government may think it simple to defeat one city's policies on climate or immigration, but let it try to defeat six hundred cities working together nationally, or six thousand acting together globally. There are more than enough reasons for municipalities to pursue collective political action on common problems, even if the legal underpinning is contested—especially if they always try first to work with, not against, national governments and international bodies, including the United Nations.[8]

Any one of these arguments in favor of urban collaboration across borders might provoke controversy, but in concert they are compelling. The rights argument aligns with the demographic, economic, and democratic arguments. Together they offer a rationale for cities to claim autonomy and collaborate with other autonomous cities, in a millennium of interdependence defined by a world without borders. Our sour national politics, like the stumbling nation-states that breed it, has for too long been preoccupied with ideological quarrels over capitalism and socialism, conservatism and progressivism, big government and no government, that no longer capture the challenges of a post-industrial, information-based, interdependent world or address the profound threats that imperil both sustainability and justice.

Finally the rights argument speaks to a fundamental way of responding to abusive authority: whether in defense of the civil rights of African-Americans, attacking the "illegitimate authority" behind the war in Vietnam, or registering the claims of transgendered people to equal treatment, the moral potency of the concept of rights has won the day. The claims cities make today to protect the welfare of their citizens gain their power from their rightfulness, which is also their righteousness. In the spirit of rights, if cities are impeded from defending their communities' sustainability, they must take to their own streets and fight for what is theirs. Mayors must manifest the rights of cities, fight for those rights, and if necessary go to court or jail for them. The way ahead is likely to be smoother than this implies, and the relationship between cities and national governments likely to be complementary rather than antagonistic. But the jurisdiction and resources cities require in order to make good on their responsibilities are not a gift handed down at the discretion of states—they are a right that cities must claim.

The struggle for urban rights will succeed only when cities join in a common effort—associated in networks that stretch across the borders of nations whose traditional sovereignty is in default. To many observers, the claims made by local governments in the name of universal rights will sound peculiar, even nonsensical. But there has been an ironic shift in the valence of national and local governing bodies with respect to universal rights. Where once nations advocated for civic, social, and economic rights while supposedly parochial states and cities opposed them, more often nowadays it is cities that promote universal claims concerning gender rights, gay marriage, higher wages, and environmental sustainability—cosmopolitanism in its true meaning—while states grow small-minded, mean, and parochial. The salient division today throughout the world is between the party of urbanity, embodying values such as creativity, imagination, pluralism, entrepreneurship, liberty, and social justice; and the party of parochialism, whether it is parochialism of the land (*La France profonde* or the Tea Party), parochialism of ethnic and religious tribes, or even the parochialism of nation-states trapped inside fading national borders.

This division can lead us to despair about national politics when a climate-denying purveyor of fear and hate like Donald Trump wins the presidency, or when xenophobic majorities run rampant in Austria, Hungary, Poland, and even the Philippines. Yet on the other side of the division are inspiring urban stories in which Londoners elect the son of a Pakistan-born bus driver (Sadiq Khan) by a huge majority as their mayor; Rotterdam residents choose a Moroccan-born Muslim (Ahmed Aboutaleb) as their Bürgermeister not many years after angry fellow citizens were calling for the blood of Muslims; and a mixed-race Englishman (Marvin Rees) becomes mayor of Bristol, once a leading port city in Britain's eighteenth-century slave trade; and meanwhile, a capital of the American South elects a brilliant African-American (Kasim Reed) to city hall while Newark, New Jersey, sends a former African-American mayor to the U.S. Senate (Cory Booker) and elects a successor, Ras J. Baraka, who is the son of the great playwright Amiri Baraka (Leroi Jones). In each case, it is a city that manifests the inclusive spirit of citizenship while nations wallow in insular nationality and competitive ethnicity. Now imagine citizens mobilizing their cities around a cosmopolitan party that is serious about implementing the Sustainable Development Goals, in whose name nations raise their voices but otherwise do little. Richard Schragger has it right: in order "to come to terms with the exercise of urban democratic power . . . a political movement is what is required."[9]

An urban political movement is not a pipe dream: in the 2016 municipal elections in Beirut, nearly 40 percent of the population abandoned tribal voting and factional parties like Hezbollah, the Sunni Future Movement, and various Christian warlord groups to vote for Beirut Madinati, or "Beirut, My City." Madinati grew from the bottom up as a result of anger over a garbage strike and an earlier "You Stink!" campaign (garbage is one of the things citizens actually care about). A winner-take-all system denied the new Madinati movement representation in city government, but given that it almost toppled the old government even as it "refused help from any veteran politicians, and ran a slate of candidates evenly divided between men and women, Muslims and Christians," as reported in the *New York Times,* it was a powerful symbolic victory for the cosmopolitan ideal incarnated as practical political struggle.[10]

Cities and their citizens, in short, have rights, and these rights allow them to claim jurisdiction in addressing climate change. Their ability to succeed where states are struggling will depend on more than exploiting the vacuum left by the sovereign default. It will require that they put flesh on the bones of right, and muscle behind their technical know-how. It will be about politics, not science—democratic politics.

2
The Devolution Revolution and the Politics of COP 21

Even nominally democratic states have struggled as they age to remain globally relevant under the weight of their ever less effective sovereignty. Over two decades they repeatedly tried—and failed—to whip themselves into action on climate. This makes the success of the United Nations Conference of the Parties in Paris in December 2015, where 195 nations committed to a historic agreement committing them to keeping the rise in temperature below two degrees Celsius, more than welcome.[1] (The annual meeting of these nations is officially known as the Conference of the Parties, or COP, initiated under the United Nations Framework Convention on Climate Change of 1992. The conference held in Paris in 2015 was the twenty-first annual meeting—so, COP 21.) But even in Paris, it was 700 participating cities that put the script of the conference into bold italics. The agreement itself gave praise to "a remarkable groundswell of climate action by cities and regions, business and civil society," and its global Action Agenda specifically mentioned the engagement of "over 7,000 cities, including the most vulnerable to climate change, from over 100 countries with a combined population with one and a quarter billion people and around 32% of global GDP."[2]

Cities had already met the previous summer in several urban assemblies hosted by the mayors Michael Bloomberg of New York and

Anne Hidalgo of Paris, and then by Pope Francis in Rome. These conclaves insisted on the need for action and prepared cities to lobby the national governments that would meet in Paris at the end of the year. They were in effect asserting their right to act on the question of human survivability in support of—and, if necessary, in place of—COP's national members. The Paris Agreement was an unprecedented commitment by the world's nations not only to keep global temperature rise this century well below 2 degrees Celsius but to make "efforts to limit the temperature increase even further to 1.5 degrees Celsius above pre-industrial levels." After so many years of inertia, there was ardent celebration. Christiana Figueres, executive secretary of the U.N. Framework Convention on Climate Change, was galvanized: "One planet, one chance to get it right and we did it in Paris. We have made history together. It is an agreement of conviction. It is an agreement of solidarity with the most vulnerable. It is an agreement of long-term vision." President François Hollande of France told the delegates assembled at Le Bourget that with such "an ambitious agreement, a binding agreement, a universal agreement," the delegates could feel "proud to stand before your children and grandchildren." And Secretary General Ban Ki-moon gave the accord the U.N.'s seal of approval, observing: "We have entered a new era of global cooperation on one of the most complex issues ever to confront humanity. For the first time, every country in the world has pledged to curb emissions, strengthen resilience and join in common cause to take common climate action. . . . This is a resounding success for multilateralism."

Yet the achievement was not without blemishes. Like all agreements that must attract the support of a broad swath of states, the Paris accord lacked specificity. Jeffrey Sachs complained it was weak on technology, in particular new technologies of carbon capture, while colleagues in the Deep Decarbonization Pathways Project warned that the accord was "failing to plan far enough ahead," and needed to anticipate goals for 2050, rather than 2030. And, to stand a chance of succeeding, cities would have to be involved. "It is mayors and city councilors and governors who are most determined to move forward," as the pledge they signed in Paris makes clear; and through building codes, vehicle purchases (electric fleets), and other measures they can make a difference.[3]

The agreement appeared specific in its appeal to a limit of two degrees Celsius in global temperatures, but the reference to something "well below" two degrees was a good deal less clear. It was necessary, however, because most climate scientists had already declared that two degrees would not be enough to mitigate ocean rise and extreme weather or repel the assault on planetary boundaries. More disturbingly, the parties could not agree on common and commensurable efforts to pursue common remedies; instead they settled for "intended nationally determined contributions." This left it to individual states to decide what they would do and when they would begin and how they would monitor and measure their progress. The United States, for example, pledged to reduce its greenhouse gas emissions 26 to 28 percent by 2025 from its 2005 levels. But the Obama administration's pledge was almost immediately challenged by coal producers, who sued over the president's use of executive orders in dealing with coal mines. In June 2016 a federal court in Wyoming struck down the Obama administration's right to regulate fracking on federal land. Even without the intervention of courts, whether future administrations live up to the pledge obviously will depend on who is in charge.

The question of monitoring and measuring emissions is obviously a key to a successful agreement. There are new technologies coming on line that may allow neutral observers to verify self-reporting by nations. By 2030, a system of satellites carrying greenhouse gas sensors is feasible.

"If built, the system could verify or cast doubt on emission reports from the 196 member states of the United Nations Framework Convention on Climate Change," helping to keep members honest.[4] *If* built. The $5 billion price tag is hardly reassuring. Who will pay for it? Politics, not science, once again is the barrier.

Consequently, it may be cities that will pick up the slack. ICLEI, the C40 Cities, and the World Resources Institute have collaborated in a joint project to establish a Greenhouse Gas Protocol, a global reporting standard for community-scale inventories. There are more than thirty-five pilot cities including Belo Horizonte in Brazil and Rajkot along with seven others in India. The protocol underpins the Compact of Mayors climate aims and contributes to the **carbon*n*** Climate Registry (cCR), which is the world's largest reporting platform on climate actions and commitments.[5]

At the national level, arrangements are messier. The pragmatic acknowledgment of the need to tailor solutions to national cultures means that all participating nations are to "submit adaptation communications, in which they may detail their adaptation priorities, support needs and plans" on a schedule largely of their own making. This assures buy-in but is hardly a formula for guaranteed progress. The plans detailing each nation's so-called nationally determined contributions (NDCs) are due only every five years, with the first "global stocktake" assessing collective progress not scheduled until 2023, and then every five years after that. This hardly seems a foundation for trust. China is already a laggard, having reported only sporadically in the past, despite having agreed in 1992 to report every five years. Its pledge to peak its emissions in 2030 will be hard to monitor. The generous stocktake intervals offer nations a very forgiving schedule, during which climate change may simply run away from slow moving ameliorative steps.

Even after nations opt in with their nationally determined contributions, they must still assure the constitutional validity of their plans by political and legal criteria that differ radically from one state to the next. Just how controversial this can be was seen almost immediately in three of the largest states signing the accord—the United States, China (whose leaders had made a tremendous impression when they pledged to bilateral climate actions well before the Paris meetings), and Brazil, perhaps the most important Global South nation to sign on.

In the United States, President Obama's good intentions were almost immediately compromised when the Supreme Court temporarily blocked his regulations curbing greenhouse gas pollution from coal-fired and other power plants. The court's action, rooted in the horizontal separation of powers, arose out of objections by twenty-nine states to the federal curbs, which they felt violated the vertical separation of powers. Since the COP agreement is unlikely to survive without the United States delivering on its promises, the agreement's success may turn on the whims of a group of American states using the courts to block the "nationally determined contribution" promised by the federal government. Ironically, the rights of states under federalism have become the instrument by which the environmental aspirations of cities and the national government are thwarted.

In China, the question is whether the government will expect reciprocity from the United States or, as some hope, the administration of Xi Jinping will see the accord as benefiting both its environment and its economy regardless of what the U.S. does.[6] As it is, China has not pledged to reduce emissions at all, only to stop them from rising by 2030. The agreement allows members to slow down the rise in emissions rather than demanding an absolute reduction; this is likely to be a feature of quite a few nationally determined contributions. It calls for nations to "peak their emissions as soon as possible and continue to submit national climate action plans that detail their future objectives to address climate change." There is no forum for assessing the ambitiousness or adequacy of these action plans, nor any agreed upon way to measure and monitor them—even though there are technologies available, as we have seen.

Regime change can also present problems. The good intentions of the Brazilian government turned on the viability of President Dilma Rousseff, who had committed her government to reduce emissions 43 percent from 2005 levels by the year 2030. This is not lightning speed, but it would represent serious progress. However ingenuous the pledge, Rousseff's impeachment as part of a broad corruption inquiry and the much more conservative views of her vice president, who succeeded her but shared few of her goals, undid Brazil's climate commitment. Similarly, even if it overcomes the current court challenge, the American commitment to the agreement will not survive a single day beyond the inauguration of Donald J. Trump. Or even a Democrat with ties to the fossil fuel industry.

In the end, even the not terribly ambitious COP 21 will depend on monitoring, verification, and enforcement, all of which the accord treats with a diplomatic touch too delicate to inspire much confidence. Governments are asked to self-monitor and self-report. So as not to scare off participants worried about the economic costs of decarbonization, the agreement tiptoes around enforcement. Its language goes out of its way to assure nations that if they cannot live up to their promises, the consequences will be nonpunitive. Nonpunitive enforcement sounds like an oxymoron. Even with the best intentions, nations (and cities too) can self-report misleadingly. Melbourne made the impressive claim not long ago that it had achieved carbon neutrality—a zero carbon footprint. It turned out, however, that "Melbourne" referred only to a small inner city region of not much more than 100,000 people, and that the calculation of carbon neutrality turned on carbon offsets the city had purchased, which does not seem an encouraging approach to global decarbonization.

For all of the agreement's brave ambitions, it is pocked with concessions to convenience that give it a Swiss-cheese feel: non-enforceable self-selected nationally determined contributions that depend on the integrity, longevity, and constitutional authority of national governments to undertake their delivery over decades. A leader out of office? No contribution. A government's authority challenged? Its intentions abruptly irrelevant. A failure to meet promises? Toothless scolding. Under such circumstances, the "powerful signal" that Christiana Figueres thinks the Paris agreement sent to "thousands of cities, regions, businesses and citizens across the world" may have been a signal less of hope than of weakness, a warning that nation-states are unlikely to deliver on their promises. If so, it will be up to cities and regions, and their frustrated citizens, to get the job done. If we truly want to stop climate change, cities are going to have to do it. Cities means us.

Another way to say this is that a successful struggle against climate change will depend on a successful revolution in urban empowerment. To make good on COP 21's promise, cities will have to grow sufficiently impatient with well-intentioned but ineffective nations, take charge, and launch a quiet revolution. Call it a devolution revolution. It is already evident in the new take-charge spirit that many city leaders are demonstrating. The emerging urbanocracy—governance by cities—is grounded in a novel approach to devolution that goes against the spirit of centralization and top-down rule that has dominated governance since social contract theory and its offspring popular sovereignty produced the French Revolution and Napoleon, and since the Articles of Confederation were succeeded by the United States Constitution. Inspired by horizontal democracy, this new spirit is reversing the centripetal impulse in governance and has been both a cause and a consequence of cities' irresistible rise.[7]

This quite astonishing devolution revolution now under way is a direct consequence of the modification of the social contract. A new recognition of the bottom-up sovereignty of citizens is already apparent in countries from Italy and France to China and the United States. It recently received a kind of official certification from the newly reelected Conservative government in the United Kingdom, formerly viewed as an archetypally "unitary" top-down political regime, with a built-in disdain for local government. Yet in the spring of 2015, chancellor of the exchequer George Osborne pronounced the "old model" of running everything from Westminster "broken" and promised a "Cities Devolution Bill." In the 2015 Queen's Speech to Parliament, Prime Minister David Cameron's government proposed a "revolution in how we govern England" that will "deliver radical devolution to the great cities of England," giving them the "levers to grow their local economy."

We can only assume the Tory government was not declaring itself a federalist regime, and that it anticipates a warm and complementary relationship between central and local government that still leaves cities in the shadow of 10 Downing Street. But tensions will inevitably arise as cities flex new muscles and begin to demand the jurisdictional authority and financial resources appropriate to the responsibilities being handed to them. In remarks given at the opening of a Singapore-sponsored World Cities Summit in New York in the spring of 2015, Mayor Bill de Blasio deftly noted that cities and national governments do not always see eye to eye, that the former's responsibilities often arise out of the latter's dysfunction, and that when national governments fail to act on crucial issues like climate, cities must and will.

Devolution of power to cities has given rise to more aggressive urban policy making in a number of nations, as well as in the European Community as a whole. It is as much fact as aspiration, having been embraced by governments desperate to see action in a world paralyzed by ideology, and it represents a powerful rebuke to national political parties wedded to neoliberal strategies of privatization and marketization as political cure-alls—partly because no one fears local government as a seat of "big government" oligarchy. And it is why (former) mayors like George Ferguson of Bristol and Michael Bloomberg of New York have stood and won elections as independents rather than traditional party partisans. Indeed, devolution is helping bring an end to the Reagan-Thatcher era of market fundamentalism and the myth that private markets do everything better than public governments. Public power is becoming localized without being privatized, and thus made more, rather than less, transparent, accountable, and democratic.

In George Osborne's description, when people feel "remote from the decisions that affect their lives," it's "not good for our prosperity or our democracy." We know there has been a sea change (in a world of rising seas) when a Tory government shifts away from a policy of privatizing power, which renders it undemocratic, to a policy of devolving power from central authorities to local but still public, democratically elected authorities. This might be seen as the bright side of the Brexit revolt against central government elites. Where the implicit tendency (and sometimes the explicit aim) of privatization was to undermine public goods and weaken government and thus democracy, the aim of devolution is to enhance public goods and strengthen democracy. That is exactly what the new urbanocracy achieves, although it often leaves the fight for jurisdiction and resources to future political struggles of the kind the Global Parliament of Mayors is designed to wage.

The quarrel is no longer with democracy. It is with the ever less viable nation-state and its cumbersome bureaucratic establishment, against which noisy populists on both sides of the Atlantic have been rebelling and whose vices more sober voices have been excoriating. The nation-state that was once democracy's proud guarantor now puts it at risk. Even before the end of World War II, Jean Monnet, the visionary of the European Community, described the sovereign central state as too big for participation, which was local, and too small for power, which was global. Decentralizing power and enhancing the public authority of municipalities is the first step to reasserting participation and accountability on a municipal scale conducive to democracy. Monnet's conundrum also requires a second, even more necessary step: cooperation among networked municipalities and a deployment of collective local power that will establish a democratic and public counterweight to private global power. This is exactly the purpose of urban collaboration, which, I will suggest below, can be facilitated by urban networks and the new Global Parliament of Mayors, a body that gives common urban voice a global megaphone, and common urban action a global platform.

The rise of cities is hardly news. Social scientists and judicious observers have been writing about it for decades, starting with classic thinkers such as Manuel Castells, Jane Jacobs, and Le Corbusier, and including such recent contributors as Bruce Katz, Saskia Sassen, Eric Corijn, Richard Florida, Ed Glaeser, and Richard Sennett. But what is new is the idea that the authority cities wield embodies a fundamental right to autonomy and resources. It is more than a little startling that a Tory government now insists British cities should have elected mayors (most do not) and is linking economic support to that condition, and is also offering its municipalities greater authority in matters of education, finance, and other domains. The national government is not only prudently deferring to urban capabilities but is tacitly recognizing that its own incapacity endows cities' actions with legitimacy—even rightful authority.

The enthusiasm for urban autonomy has been accompanied in many places by an interest in organizing effective metro-regions. As power flows down into cities, cities expand their compass and power flows out to the suburbs, exurbs, and surrounding farmlands, forging partnerships in what become new metro-regions. These metro-regions are far more encompassing than the old walled cities of the medieval epoch. The metro-regional impulse is not, as skeptics have suggested, just a way for cities to take over their suburbs.

Prime Minister Matteo Renzi of Italy, formerly the mayor of Florence, recently introduced a constitutional reform that created nine core metro-regions to succeed Italy's traditional provinces. The new administrative infrastructure, if adopted, will be represented in a reformed Italian Senate. In France, the city of Paris has recognized that its twenty wealthy inner-city *arrondissements* must learn to live with and share the problems that lie just beyond the *peripherique* in the outlying *banlieues,* aging suburbs where today so many marginalized immigrants live isolated from the City of Light. During the mayoralty of Bertrand Delanoë (2001–2014), the inner city of 2.5 million developed ideas for a Grand Metropole that would encompass not only the first ring of suburbs (the three departments around the peripherique), where as many as five million people resided, but four additional departments in the second suburban ring with three million more—a total of some ten and a half million in a Greater Parisian Region almost as extensive as the traditional Île-de-France. While this ambitious plan ultimately failed, in part because of controversies over taxation, the new city government of Mayor Anne Hidalgo is pushing for a more modest *grande metropole Paris* plan focused on the three departments nearest Paris. One way or another, the regionalization of Paris has acquired a momentum that can ameliorate the sharp division between rich inner city and poorer immigrant suburbs.

Even in China's top-down unitary state, where the Communist Party is ever alert for signs of national disintegration, cities are being given greater local authority in economic and environmental affairs. It cannot be called devolution, but the party already endows the mega-cities Beijing and Shanghai and their mayors with special competences. For the mayor of Beijing, particulate pollution from coal is a problem that will not await the lethargic response of a national Communist Party. Other mayors are held on a much shorter leash and are appointed and removed at the party's will (or whim). Yet they can exert significant leadership, as Mayor Geng Yanbo of Datong did with his attempt to rebuild the city's ancient wall as a tourist attraction, until he was removed by nervous party officials worried by his popularity and ability to get things done with an atypical engagement by citizens. His remarkable tenure is depicted in a compelling film, *The Chinese Mayor,* by the Chinese director Zhou Hao. And Hong Kong, impelled in part by its unique historical circumstances, has managed to mount a resistance to the mainland government's overweening hegemony.

Devolution has been a more natural feature of federalist regimes, for which the model is the United States, where the recent incapacity of the national government has boosted the prominence of American cities. As observers such as Bruce Katz and Parag Khanna point out, local economic topographies often bear little resemblance to the topography of political boundaries. Cities that start out compensating for federal insufficiency, acting in place of an ideologically paralyzed national government in domains such as gun control, fracking, civil rights (gay marriage, LGBT rights), and the minimum wage are already beginning to view the acquisition of the jurisdiction and resources as a necessary condition for discharging their responsibilities. Empowerment by default becomes empowerment by right. The devolution revolution is birthing a world in which cities are becoming indispensable and perhaps even quasi-sovereign players in addressing the planet's most urgent problems. Foremost among these is climate change.

3

Climate Change in the Anthropocene

Our new epoch has witnessed a profound shift in human life and civilization, from stasis and continuity to change and disruption. After millennia of inertia, in which the scope and activities of humanity remained in delicate balance with the material planet, we have just recently (and abruptly) gained the capacity for worldwide disequilibrium. The Anthropocene is the *human* epoch, an era in which, over the last two hundred years since the coming of the machine, our species has become the major factor driving planetary change, rather than accommodating itself to planetary boundaries. By establishing a foundation first in machines and then in information technology for economic and social progress, humans have altered the natural order in favor of triumphant artifice.

Earlier epochs during and after the Ice Age were named for their natural geological character, such as Holocene and Pleistocene. The Anthropocene, referencing a reversal of the traditional natural order in which humans were an integral part of nature, denotes the novel ascent of our species to dominion over nature and the imprint it will inevitably leave on the geological record. In the political context, it reflects not just progressive civilization on the make but the costs of civilizational progress. These costs have put the sustainability of planet Earth at risk. The term *Anthropocene* captures not only the essence of post-industrial modernity but also its dialectic, the contradictory face of industrial civilization that philosophers have associated with the dialectic of enlightenment.[1]

As a technical matter—and a great many scientists, economists, and experts regard it primarily as a technical matter—the consequences of the Anthropocene are clear: our efforts at mastery have disrupted the planet's ecological balance and thus imperiled the human future. Our progress is secured by processes that encroach on our natural world's "planetary boundaries"—limits that defined the conditions necessary for humankind to survive its encounter with natural forces, of which climate change is only one. Together, the boundaries define a "safe operating space for humanity." As first described in 2009, they include climate change, ocean acidification, stratospheric ozone depletion, the nitrogen cycle, the phosphorus cycle, global freshwater use, change in land use, biodiversity loss, atmospheric aerosol loading, and chemical pollution.[2]

The boundaries suggest how interdependent our manifold crises are. Overstepping the climate boundary contributes to ocean acidification, the lowering of water tables, and the acceleration of biodiversity loss. The impact of ocean acidification on the massive loss of coral reefs in Australia, for example, is in turn affecting fisheries and sea creature biodiversity, developments dramatic enough to attract widespread media coverage.[3] The fragility of these boundaries and the results of our ignoring them are manifested first of all as climate change, but also as ocean acidification, ozone and groundwater depletion, changes to the nitrogen and phosphorus cycles caused by runoff from fertilized land, and not least of all, biodiversity loss. Together these ominous features of the Anthropocene, closely linked to human economic and technological advances, have brought us in just a few centuries to the brink of both unprecedented progress and a natural catastrophe that endangers our civilization, our species, and other species as well. We seem not to have absorbed the lessons taught by the Greek story of Prometheus. Modern hubris lies in a new theft of fire associated with the burning of fossil fuels, pushing our civilization forward at once to greatness and to the brink.

Technically, there is little mystery about how such Promethean trends are to be addressed. The title of a recent book by Jeffrey Sachs, *The Age of Sustainable Development,* appears to suggest that we already have entered that new age and attained the solutions we need—or that we will shortly, just as soon as we apply what we know to fix what's gone wrong.[4] We need only acknowledge and implement the United Nations Sustainable Development Goals, born of the U.N.'s earlier Millennium Development Goals; we need only accept the logic and aspirations of the United Nations' COP, which since its beginnings in Berlin in 1995 has pursued a climate change protocol that enables (compels) nations to prevent a rise in global temperature of more than two degrees, or an increase in atmospheric carbon dioxide to more than 350 parts per million. With the COP 21 accord reached in Paris at the end of 2015, this aim was realized. So now we need only realize any number of further urban-inspired initiatives from the Compact of Mayors and the Urban Agenda to the C40 Cities Climate Leadership Group criteria devised jointly by scientists and city leaders to promote long-term decarbonization. Do all this, and the problem will be solved.

There is no good reason why all those two-sided technologies of internal combustion engines that engendered the climate crisis should not be used to fix it. There is no reason why all those innovations of information technology and digital engineering that bring indirect pressure to bear on planetary boundaries should not be turned toward sustainability. Physician heal thyself! Let technology repair the natural world it has put at risk. Promethean fire kills by burning but sustains by warming. Fight fire with fire.

From the darker mood of technological anxiety can be derived a welcome optimism about solutions. It is always a mistake to underestimate what technology can do to undo what technology has done. We will look carefully at what is possible on the road to decarbonization in Part II. Greenhouse gases like methane and carbon dioxide can be captured or reburied or even converted to energy. Garbage waste can be used to generate heating and biomass fuel rather than being added to methane-leaking waste piles and space-consuming landfills—or being exported to countries too poor to resist the profits that come with being another nation's garbage dump. We can adopt electric vehicles, commercial and private, some now self-driving, and of course would need no vehicles at all in walkable cities; we can employ low-impedance transmission wires, local generation of energy, garden towns, water management schemes, LEED insulation standards, resilience programs, aboveground mining of chemical elements, wastes, and other "accessible elemental resources of our future" that are "largely stored aboveground in the familiar objects of our daily lives."[5] We can even build wooden skyscrapers, like the thousand-foot beauty proposed for the Barbican in London. The sky is quite literally the limit. Technical fixes are constrained only by imagination. And time.

Time, however, is a problem: it walls in imagination. The fix may come too late. Or be on time but call for corrections that bring it on line too late. It may bring unanticipated consequences. New energy technology can drive down the price of the old carbon technology, giving it a new economic lease on life. Wooden skyscrapers create new fire hazards even as they please environmentalists. Electrical vehicles increase demand on electric power systems, pushing carbon use into the background where its pernicious effects are less visible. (Your wonderfully green electric car may actually be running on coal-fired electric power.) The fix can become the problem, as when an unexpected shift in the wind pushes the flames from an artificially set firebreak the wrong way, accelerating the fire it was meant to slow.

Too often an imagination-driven optimism meant to turn technology against its own depredations can become a naive technozealotry that makes things worse. The appeal to "smart" solutions (smart technology will fix everything) actually requires a sophisticated appreciation of technology's limits—and a careful lookout for unintended consequences. Smart parking meters aim at relieving traffic congestion and saving gas by helping drivers get more quickly to an empty parking space; but they can backfire and increase traffic if they encourage more people to drive into town rather than take public transport. And smart technology can be put to stupid or even malevolent ends. The Nazi death camps ("smart death camps"?) incorporated new technologies that allowed killers to gas, murder, and incinerate human bodies more "efficiently" than ever before. Edwin Black showed how IBM's state-of-the-art information systems at the time helped engineer the genocidal process.[6] It is the purpose to which technology is put that creates a "smart" security program (do endless cameras enhance security surveillance at the cost of annihilating privacy?), or produces a "smart city" (ghettos are actually quite smart—that is to say, efficient—if the aim is to maintain a permanent urban underclass).

Even as important an idea as resilience (more on this below) can have two faces. When used to prepare cities for natural catastrophes and address the inevitable costs of preemptive strategies that fail, they are a vital tool in cities' survival toolkit.[7] Human frailty and error must be built into every program of reform and change, sustainability included. Yet in the hands of cynical players, like some carbon energy monopolies, the pursuit of resilience can become a rationale for inaction. By lumping together natural disasters with human-induced ones, the call for resilience risks relieving us of responsibility for the outcomes of our own bad policies. As long as we can "fix" what went wrong later, there is no wrongdoer who cannot be forgiven. Being well prepared for the worst or ready to deal with the high costs of convenient choices made out of self-interest relieves us of the need to make more prudent choices when choices are still there to be made.[8] Prepare now for the urban flooding that is sure to come later on if we persist in our bad habits, and we can stop worrying about the bad habits, or do anything at all to preempt their consequences.

For natural scientists and those who count on the rationality of both science and the species that made it, sustainability can feel like something just beyond the horizon, approaching fast and asking only that we embrace it, prudently apply its lessons, and do the things science tells us to do based on what we already know. This attractive but illusory approach settles on *knowing* as the key to what we must do. Knowing and willing seem almost synonymous: we need only listen to experts and turn their knowledge into action. For rationalists, ignorance is action's only enemy, and ignorance is always susceptible to enlightenment: the only real task is to educate and enlighten those who don't know. Explain the science to them. Show them the facts.

This is not how the world works, though. There is a kind of Platonic hubris in climate scientists who harbor the certainty of philosopher-kings that knowing what must be done means it will be done. In a world of reason, maybe. But the world of politics is not a world of reason. In the world of politics, facts are mute and understanding is not action. To will action entails more than just knowing.

4

The Facts Are Mute, Money Talks

So here's the colossal hitch. Dealing with climate change and planetary boundaries is not about what we know, it is about what we do—what we will politically to do. Knowing what to do is not the same as doing it. Natural scientists tend to believe in rationality and a natural order of things, and hence in the connection between understanding the natural self-evidence of solutions and willing their implementation. Once we know what needs to be done, why would we not do it? That's why those who would combat climate change place so much emphasis on sharing information and getting the facts straight. Facts speak for themselves!

Except when they don't. Despite the obvious necessity to curb greenhouse gas emissions and reel in the carbon economy over the past several decades, carbon dioxide (CO_2) in the atmosphere has climbed above 400 parts per million, well over what scientists consider the maximum safe concentration, 350 ppm. Nation-states know this. They have known for decades exactly what the facts are, and have done nothing. Even with the progress made in the twenty-first COP meeting in Paris, we remain on the road to irreversible warming of the atmosphere, with all its dire environmental consequences. The facts may speak for themselves, but in the domain of politics and collective human action, they are mute, without influence on outcomes. Nature doesn't bargain, but in politics it's all about the art of the deal.

Forget the facts—it's money that talks. And power that brays. In politics, brute force, messy and weighty, is worth a thousand precise words. Those engaged in the practice of politics and the political scientists who study it will insist to the rationalists that whether we actually manage to enter an age of sustainable development will depend not on knowing the right thing to do but on summoning the will to do it. That is to say, it will depend as much on our interests as on our values, and in the struggle among competing values, on which ones prevail. And it will depend on whether we reckon the worth of our interests against the worth of our values in a private ledger or a public one.

The bottom line is that getting what's good for us (sustainability) hinges on how much power we can deploy and in whose interests. Scientists will say with conviction—and they are right—that "there are no technical or economic barriers to achieving sustainability."[1] The technical fixes for making development sustainable are straightforward and compelling: shift to alternative energy, reduce carbon use. A carbon tax, maybe? (Something everyone once agreed on!) Eliminate waste, rewire the grid with low-impedance cable, insulate buildings better, push bike sharing, recycle, improve efficiency, green the economy. Do solar, do wind, do geothermal, do tidal—do them all!

The economic facts are even more compelling: it would require spending only 2 or 3 percent of world GDP to secure sustainability; less than 5 percent of world GDP to put an end to poverty. We have the capacity to achieve sustainability and end poverty, now, at marginal cost and with huge long-term benefits. But so what? Who cares? Or better, who cares among those who "matter" politically? Among those who have skin in the game, interests to be protected, and the power to protect or advance them?

So although there may be no technical or economic barriers to solving the climate crisis and the related issues of planetary boundaries, there are myriad political barriers to change, and many of them look unmovable. The political game does not play out in a vacuum where the only players are the good guys who have figured out how to do it and a few ignorant ones who don't yet get it. There are powerful interest groups perfectly knowledgeable about the consequences of greenhouse gases that nonetheless are unwilling to pay the cost of decarbonization if it affects their profits. There are vested private stakeholders who understand the causes of (and may even deplore) poverty but benefit from poverty, exploit it as ruthlessly as they can, and engage in activities that have the "side effect" of increasing inequality.

Market players often give personally to charities like Mercy Corps and Doctors Without Borders even as they support systemic political and economic arrangements that make charity at best marginal to real change. By the same logic, there are private-market players who see in alternative energy not just a curb on carbon emissions but a curb on profits. Many of these people understand the science of climate change perfectly well; they are not deniers and don't need to be. They are not ignorant, certainly not stupid; they are simply driven by interests at odds with sustainability or climate justice. What's good for them and what's good for the planet are just not the same. They may begin by making scientific arguments rooted in expensive and politically improbable fixes that allow oil and even coal to be extracted from the earth until every last ounce of carbon is drilled, mined, and burned. This is Samuel Thernstrom's strategy in his paean to "enhanced oil recovery," which he says will usher in the "next shale revolution."[2] But rational arguments about technology are not really the key. In a speech to the Western Energy Alliance's annual meeting in 2014, an influential consultant spoke of the need to combat "Big Green Radicals" by "exposing environmental groups." Not with facts or arguments, but through PR campaigns emphasizing what he called "FLAGS": "fear, love, anger, greed, and sympathy"; think of it, he concluded, "as an endless war."[3]

The problem raised by lobbyists and shills for big oil funding politicians committed to deregulation, or by a Shell oil rig heading to drill in the Arctic despite the environmental risks of Arctic drilling, is not technical, it is political.[4] It is not about explaining to those who profit from carbon what's wrong, or about teaching them what's right. It is getting them to see politically that what may be "right" for them as carbon corporation stakeholders is wrong for them as citizens of an interdependent community called Earth. What is required is political persuasion—maybe regulations or tax codes that realign their economic incentives—to bring their behavior into accord with public rather than private interests. Which is to say, it is not about science, it is about politics, about interest, about power. It is about how power defines the "we" we refer to when we talk about what "we" really want. It is about which "we" holds power—and in whose interests it acts: whites or non-whites? men or women? The one percent or the ninety-nine percent? Some of us or all? Those who can flee inland when the ocean comes to claim our cities, or those who will drown?

The focus of politics is interests, not goods. Harold Lasswell's famously dismal text on power contained this message in its very title, *Politics: Who Gets What, When, and How.*[5] The pertinent problems are problems of governance, institutions, and democracy. If Jeffrey Sachs were a political scientist and a student of brute force and violence, rather than a rational economist and a student of natural science, he might have called his important book *The Age of Unsustainable Development,* with the new subtitle "Can We Do Anything About It Given That We Know the Remedies but Refuse to Apply Them?"

Human beings are not rational. If they were, not only climate change but poverty would have been addressed long ago. And injustice and war. Has there ever been a rational war? Once in a while, against a genocidal aggressor, a necessary war perhaps; but rational? Never. The big questions are not about rationality. They're simply about politics. Which is anything but simple. Or rational.

5
Privatization and Market Fundamentalism

Like it or not, we live in an age of stubbornly unsustainable policies pursued by stubbornly interest-bound political and economic institutions underwritten by stubbornly private-market fundamentalists who stubbornly privilege wealth over equality and profit over sustainability. The obstructionists preventing national governments from governing do not act out of ignorance or in denial of the facts. They prefer to indulge in the tactics of the so-called merchants of doubt who affect to be healthy skeptics (aren't scientists supposed to ask questions!?). This sly tactic of pseudo-scientific skepticism was pioneered by the tobacco lobby in the 1950s and helped it hold out for decades against irrefutable medical and epidemiological evidence showing the correlation between smoking and cancer.[1] The lobby did not deny outright that there might be a correlation; instead it impersonated prudent interlocutors by casting doubt in the name of science. The tobacco lobby then, like the carbon energy lobby today, was hardly motivated by a love of science and its logic of falsification (doubt). The climate skeptics use doubt to promote self-interest—a disposition at odds with the kinds of facts that might prove inconvenient to their interests.

Under the euphemism of "private goods" and "market values," the self-interested simply shoved aside public goods and common values. They were indifferent to the vanished idea of "commonwealth" that once animated free republics. In fact, we know the cost of climate change to public goods (or can estimate it); we know the value of decarbonization in pursuing sustainability. That value is in the future, however, and not salient enough in the present to moderate the positions of oil industry executives liberated by the *Citizens United* decision, or politicians-for-hire, or ordinary citizen-consumers more in love with big cars and air conditioners than the welfare of their grandchildren. Such are the intransigent power realities that will dictate the outcome of the battle for sustainability in the absence of a successful political struggle to defeat them.

It may seem to some that if it is about politics and not science, things will be easier. We know what to do, we need only find the political will to do it. In fact, politics makes things harder. Much harder. The daunting political reality is that many players in the Anthropocene are on the wrong side of the dialectic: they do not *want* to address the overstepping of planetary boundaries, even though those boundaries stand before them as bright red lines on a brown field stretching to the horizon. Their interests are simply at odds with the global human good. Private desire trumps their commitment to public virtue.

To natural scientists, then, solutions are simple and clear. To political scientists, the likelihood of such solutions being implemented is doubtful. There are three particular ways in which the politics of climate change play out that make the prospects of science-based sustainability problematic.

The first is a consequence of the politics of privatization and market fundamentalism. In the last thirty years, privatizing trends have incapacitated not only the political state but democracy itself. By catalyzing capitalist triumphalism, they have removed government from the tool kit of reform and remediation, whether in the domain of climate change, economic justice, or corporate regulation.

The second factor is a consequence of democracy's corruption. Even when it manages to resist marketization, democracy often gets reduced to majority rule and private polling in a manner hostile to deliberation and thus to sustainable policies. While the majority is often critical of green policies, this is evidence not of democratic perverseness but the perversion of democracy. The trouble comes when democracy is misconstrued as nothing more than the public polling of private opinion and the uncritical registrar of majority whims.

The third political factor that has been pernicious to the struggle for sustainability is a consequence of the continuing primacy of the traditional nation-state as democracy's putative guardian. As we have seen, the sovereign default of nations no longer able to cope with the challenges of the Anthropocene and the brute facts of interdependence put the credibility of democracy itself at risk. No institutions are more fragile today than sovereign nation-states and the international organizations built on their crumbling foundation. Many nations have become dysfunctional entities, and I mean this as a description of reality, not to cast an aspersion. The most cursory look at world events in the second half of 2016—in Belgium, Brazil, Spain, Venezuela, Russia, or the United States—confirms that national governments like the one in Washington are, in Thomas Friedman's colloquialism, "stuck." "There is," he laments, "an overwhelming sense of 'stuckness'" that helps explain the popularity of anti-establishment populist politicians breaking out in one nation after another.[2] A period of rapid climate change is a very bad time for politics to be stuck, or worse still, in the hands of demagogues. One antonym for stuck is *unglued,* hardly an improvement.

These three aspects of our current politics are worth some attention. The first concerns how market ideology has skewed the traditional balance between public and private, between the state's common goods and the market's private interests, in ways that obstruct government action on climate. For almost forty years, what George Soros calls market fundamentalism has threatened the very idea of government: the planning, taxation, and long-term deliberative vision essential to sustainable climate policy. This is more than a matter of corruption: it is a fundamental undermining of politics by the ideological triumph of neoliberalism, of the view that government is the problem, not the solution, while the market—seen by democracy as the problem—is the solution.

In the United States, the ass-backward ideology that money cannot corrupt politics because money *is* politics was given constitutional authority by two key Supreme Court decisions. The first, *Buckley v. Valeo,* ruled that money is a form of speech; in the second, *Citizens United,* corporations—construed as legal persons—were declared free to spend as much as they pleased on political campaigns. The two decisions together validate the corrosive proposition that when corporations spend money they are for all relevant constitutional purposes to be treated and protected as persons exercising free speech. The 2016 presidential campaign along with the vacancy on the Supreme Court left by the death of Justice Antonin Scalia placed the emphasis on *Citizens United.* But the real stumbling block is *Buckley v. Valeo.* Bickering about whether corporations have a right to speech does little to challenge the far more insidious proposition that money is speech. Money was long understood to be a skewed form of power for which equal speech—free speech—was a necessary remedy. The finding that money is speech removed a primary remedy to the abuse of power.[3]

The skewing of the public-private balance toward the market, the second aspect of politics under duress, is not a reflection on the capitalist system per se. The strong critique of capitalism associated with politicians like Bernie Sanders and journalists like Naomi Klein is insufficiently dialectical. Capitalism not only remains the only economic system left standing in the post-Soviet modern world, it continues to be the most productive wealth creation system the world has ever known. But it is not very good at protecting competition—it generates monopolies that undermine not only democracy but capitalism itself—and is even worse at promoting distribution. It produces wealth, but sometimes at the expense of well-paid jobs—or any jobs at all (automation is profitably efficient). It privileges individual property and private liberty but neglects equality and is oblivious to justice.

What it does, capitalism does well enough. Those goods it does not or cannot produce, such as competition, fair distribution, jobs, equality, justice and fairness, can more efficiently be produced elsewhere. They are public goods, the responsibility of the public sector, of which government is the principal guardian. Private individuals (the aggregated "me") may be selfish and self-interested, but when they see themselves as democratic citizens (the common "we"), they are able to contain private interests (their own included) by balancing them with public goods. If brute force and money are the engine of private interest, legitimate democratic power is the steward of the public good. Individuals do for themselves; citizens do for their communities, and hence for themselves as members of the community. When markets take precedence over civil society and private wealth dominates commonwealth, the problem is not too much capitalism but too little democracy. Capitalism unleashed and markets unhinged are a function of democracy incapacitated.

Blaming capitalism is a stratagem of failed democracy. When the public sector is strong, capitalism need not be weak. Its excesses will be curbed and its contradictions contained. Individual property owners and capitalist shareholders can focus altogether appropriately on private profit and property in their capacity as private persons, knowing that government, national and local, embodying the will of those same persons when they conceive of themselves as citizens, will focus just as appropriately on justice and the common good. They will constitute themselves as civic correctives to the self-interest they manifest as individuals. Sustainability depends less on capitalism weakened than on democracy strengthened. Strong democracy makes for strong climate action. Cities are the key to democracy and hence to common action on climate change.

This is not to say that democracy is always a friend to sustainable policies. The second reason why democracy (at least in name) does not automatically lead to sustainability is that democratic governance, even when restored to its appropriate place, is not necessarily a friend to long-term deliberation. Democracy may reflect the choices of a citizen community dedicated to public goods, but their choices can be driven by short-term perspectives (cheap energy, for example, or air conditioning for all) rather than the goods of tomorrow (alternative energy or better insulation). There is democracy and there is democracy.

In an ideal world, a democratic regime representing informed citizens would move aggressively to address global warming. Protecting the future of the planet with vigorous policies should come naturally to conscientious nations responding to energized citizens. Democratic deliberation is designed precisely to help selfish individuals reformulate their interests in the language of the communities they belong to—move from short-term "me thinking" to longer-term "we thinking." At its best, democracy allows private opinion to be shaped by shared civic belief and the discipline of intersubjective ("objective" or "scientific") knowledge. Applied to climate change, the deliberative process should produce successful and sustainable environmental policies. And it is of some promise that the partisan divide has been challenged by some Republicans grounded in the kind of conservatism represented by *conservationists* such as Theodore Roosevelt.[4]

Yet this modicum of bipartisanship notwithstanding, it is pretty obvious, however, that we do not live in an ideal world. In our real world of corrupted, minimalist government dominated by money and special interests, democracy is hardly at its best. Even when government is seen as a legitimate player in regulating the market, its compass can be restricted and its sphere of influence limited. Market fundamentalism taints the notion of democracy itself. It urges citizens to act like consumers and to think their civic job is to express impulsive private preferences. Rather than seek deliberative consensus, they are encouraged to see the public good as only an aggregation of private preferences.[5] Opinion and knowledge are confounded, leaving judgment in limbo. Some voices even try to persuade us that democratic thinking itself stands in the way of expert science—popular majorities are not always friends to civic self-sacrifice. But shared ignorance and democracy are not the same thing. The journalist Andrew Sullivan, for instance, recently wrote that "democracies end when they are too democratic," and that only prudent Madisonian measures limiting democratic culture and dividing government can contain abuse.[6] From this perspective, the issue is not democracy corrupted but democracy itself.

In this antidemocratic critique, the tyranny of the majority is identical to democracy, so that when "now" trumps "later" and today takes precedence over tomorrow in ways that preempt action on climate change, that simply is what democracy does. When democracy equals mobocracy, democracies cannot by definition do anything at all about policies opposed by the mob. Edmund Burke, however, wisely treated the democratic contract not just as a contract among the living, but one that links the interests of the living with those of the dead and the unborn. This kind of intergenerational thinking can only be cultivated in a setting of prudent deliberation and civic-mindedness. Sullivan and others think this is beyond the democratic imagination, but I believe it *is* the democratic imagination incarnate.[7] It is our current present-mindedness that falls short of the democratic standard.

It is apparent that democracy—under siege from corporate capital, more beholden to the private sector than to public goods, and misread by critics as mobocracy—is more often seen as obstructing than facilitating green policies, above all on the national political stage. It does little good to restore the state to its role as sovereign regulator of the private sector, when the private sector owns the state. Or when the people are thought to comprise a sovereign mob. The word *sustainability* should push citizens out of the mindset of "now" and allow them to temper today's wants with tomorrow's responsibilities. But when citizens fail as deliberative judges of their own long-term interests, it becomes tempting to think that a benevolent tyrant with an understanding of the science will address climate change more effectively than so-called citizens; that is to say, that the Central Committee of the Chinese Communist Party is more likely to act responsibly than the U.S. Congress. How can advocates of democratic governance respond to so perverse a conclusion? Do we give up on democracy? Or on sustainability? Actually, there is a third option: reanimating democracy by devolving power to cities.

6

Political Institutions Old and New: Cities Not Nation-States

Let us focus, then, on the institutions that are democracy's instruments and guarantors. The decades-long lassitude of nations in the face of the climate crisis was made manifest with the failure of the first twenty rounds of COP meetings after the United Nations first convened them in 1994. COP 15, in Copenhagen in 2009, seemed on the verge of success, but hope waned quickly and dispirited nations lapsed back into controversy and inaction. More recent COP assemblies in Warsaw (COP 19) and Lima (COP 20) did nothing to slow the pace of warming, leaving citizens with a growing sense of what can only be called both a democracy deficit and a sovereignty default. Only at the very end of 2015 in Paris, following gallant urban leadership the previous summer by former mayor Michael Bloomberg of New York, Mayor Anne Hidalgo of Paris, and Pope Francis, did the COP 21 congress finally achieve results—not least because of the representatives of seven hundred cities attending as urban watchdogs.

The Paris agreement also rode a wave of popular political demonstrations, for which the half-million-person People's Climate March, in New York in September 2014, was the model. Pope Francis's encyclical on climate and justice, *Laudato Si'*, created an imperative that nations found it difficult to ignore, and the pope followed this widely read encyclical with a congress of cities held in the Vatican in August 2015.

The new mood was captured in feel-good essays by a formerly cynical press corps—notably Jonathan Chait's misleadingly optimistic cover essay from September 2015 in *New York* magazine, "The Planet Isn't Doomed After All: A Sudden Reversal on Climate Change." In response to the doomsday realism of the climate scientist James Hansen, who had warned that in the coming decades "forced migrations and economic collapse might make the planet ungovernable, threatening the fabric of civilization," Chait gushed that there was "not only incremental good news but transformational good news." It looked like "the good guys are starting to win."[1] We can only hope.

Yet states are held captive by money, business, and banking interests as well as a corporate media that often misleads rather than informs the public. The resulting democratic deficit makes effective climate action difficult, and at the national level nearly impossible despite agreements like the COP 21 accord.

In the United States, we have seen how the Supreme Court's *Buckley v. Valeo* and *Citizens United* decisions gave constitutional legitimacy to the tainting of democracy by money. Meanwhile, "charitable" lobbies such as ALEC (the American Legislative Exchange Council, a tax deductible organization that taxpayers unwittingly contribute to) give private power a public voice in thwarting sustainable policies. ALEC provides state legislators with free trips and access to potential campaign donors, in exchange for their support on legislation—generally drafted by ALEC—to advance the interests of its member corporations, often in narrow terms that serve specific industries. It claims to be a membership organization for legislators but is largely supported by corporate donations, including from the carbon industry.[2]

It is not, however, just a matter of democracy corrupted. National governments rooted in sovereignty, trying to shape global politics across their borders, are no longer positioned to govern effectively, either singly or in common. Top-down centralized democracy is confronting a world in which, as Friedman noted, "all top-down authority structures are being challenged," and where there is greater opportunity for pluralistic societies that can govern themselves horizontally. These conditions challenge the authority of monolithic governments that believe their societies can only be "held together top-down with an iron fist."[3] Even if democracy were less compromised than it is, even if it didn't so often seem like a rationalization for plutocracy, it would remain trapped inside the box of national sovereignty.

Hence, our third dilemma: the dilemma of bordered and blinkered independent states confronting borderless, interdependent problems. Every challenge we face today crosses borders. Climate change, terrorism, refugees in flight from genocide, civil war and economic meltdown, labor, commodity, and capital markets in turmoil, pandemics, crime, drugs, weapons of mass destruction, and the anarchy of our ubiquitous digital technology—all are global in their causes and consequences. No Chicago warming, only global warming; no Tokyo internet or Paris web, only the World Wide Web; no state-based war, but malevolent NGOs like Al Qaeda and quasi-states like ISIS, accompanied by endless civil wars. States such as Libya and Iraq have effectively ceased to exist.

We confront these brutally interdependent challenges with antiquated nation-states, wrapped in the very sovereignty and independence that leave them incapable of meeting the new perils. We have HIV without borders, war without borders, immigration without borders, a digital web without borders, but we do not have citizens without borders or democracy without borders. Who do we imagine can contain global warming without borders if there is no government without borders? On this devastating asymmetry between problems and responses turns our future. Unless we find ways to globalize democracy or to democratize globalization, humankind will be in ever greater peril.

The institutions we think of as global or international are all state-based: the United Nations and the international financial institutions associated with the global system (the World Bank, the Asian Bank, the new Asian Infrastructure Investment Bank); the World Trade Organization and the International Monetary Fund. We depend on them as international entities to help achieve solutions to cross-border problems. The U.N. secretary general's office and the European Parliament do try to assume transnational leadership in pursuit of sustainable goals. Yet the nation-state was conceived in an age of independence, where national jurisdiction circumscribed human problems, leaving them amenable to amelioration only from within. The borders delimiting state action are irrelevant to such global perils as climate change. This makes COP 21 a dubious "success."

I want to suggest, then, that we can find an answer to the dilemma of independent, bordered nation-states wrestling ineffectively with interdependent, borderless challenges, and thus to our inability to address climate change through nation-state democracy, by changing the subject. From states to cities; from prime ministers and presidents to mayors. Our most ancient and enduring political bodies—our towns and cities—offer an attractive alternative to dysfunctional nations. Let interdependent cities do globally what independent nations no longer can do: let mayors and their neighbors, the citizens of the world's cities, address climate, regulate carbon, and guarantee sustainability through cooperative action. Let mayors cool the world.

There are good reasons why cities can effect changes that nations cannot.[4] We have always been what Edward Glaeser calls "an urban species." Today more than half the world's population lives in cities; in the developed world, more than three-quarters do. China is growing new cities of more than a million at a dizzying rate. A few decades ago, Shenzhen was a town of perhaps 20,000; today it is a megacity of over

18 million. Meanwhile, burgeoning conurbations in Africa and Latin America are making New York and London look provincial. Cities, as I have mentioned, generate nearly 80 percent of GDP and 80 percent of greenhouse gases. And because they create much of the problem, they can contribute significantly to the solution, if they have sufficient resources and can act with sufficient autonomy. As Michael Bloomberg reminds us in his provocative article "City Century," cities can act more quickly than states and are less likely to be "captured or neutralized by special interest groups."[5] For mayors, reducing carbon pollution "is not an economic cost; it is a competitive necessity" that manifests the "congruence between health and economic goals."

While asserting that "the world's first Metropolitan Generation" is just now coming of age, Bloomberg notes that the city stands at the beginning of our history.[6] Human civilization was born in cities, and democracy was first nurtured in the polis. Cities are our most enduring political bodies. Rome is much older than Italy, Istanbul older than Turkey, Boston older than the United States, Damascus older than Syria. Cities are where we are born, grow up, go to school, marry, and have children; they are where we work, play, pray, grow old, and die. Concrete and palpable, they draw their existence from their concrete, organic growth rather than from boundaries drawn on a map; from the art of communal life rather the science of public administration. Cities define our essential communitarian habitat in a way nation-states cannot.

Nations are too large for participation and engagement but too small to control the global centers of power. Too big for community and association but too small for the world economy. Cities are closer to us, more human in scale, more trusted by citizens. Fewer than half of Americans trust the president or the Supreme Court and less than 10 percent trust the Congress they themselves elect, but 70 percent or more trust their mayors and municipal councilors. The same is true worldwide: local government is deemed more trustworthy than national government except in a few nations, such as China, where local government isn't local but is controlled from the center. (To the degree that they trust anyone, the Chinese appear to trust the party and the central government more than their local leaders, whom they consider impotent pawns.)

To respond effectively to climate change, we need to restore democracy to its deliberative roots in competent citizenship at the municipal plane. It is easier in the city to reinstate popular government as a domain of deliberation, accountability, and citizen participation. The neoliberal assault on "big government" has little traction in cities, where government is small and focused on sewers, schools, policing, housing, traffic, and jobs. Nations stare out suspiciously across fortified borders at neighboring countries, while tribal nativists call for higher walls and prime ministers appeal for higher defense budgets. Antagonism is the modus operandi and war is its final recourse. Cities are open and transactional, defined by trade, culture, and commerce. Nations are often in a zero-sum game: when Germany grows larger, Poland grows smaller. Yet Berlin and Warsaw can both flourish without thinking that the success of one must entail the other's failure. Indeed, their relations in trade, culture, transportation, and environmental sustainability are necessarily interdependent. Success requires cooperation.

We can address climate change, then, by talking about cities and asking that their mayors talk to one another. Environmental sustainability will be achieved when we secure sustainable democracy, and democracy is sustainable today mainly in the municipality.

Cities have an enormous potential for ecological cooperation, engaging their citizens directly in climate action (through, for example, pedestrian zones, recycling, and mass transport) even as they act on a global scale through collective action. They are already actively seeking sustainability across national borders through urban networks such as the C40 Cities, ICLEI, and Energy Cities Europe. These networks, undergirded by larger, less specialized associations like the U.S. Conference of Mayors and the United Cities and Local Governments network, are not very well known. But they are hugely effective intercity associations allowing cooperating cities to do what nations have failed to do (the details will come later).

If presidents and prime ministers cannot summon the will to work for a sustainable planet, or even live up to the modest agreements they so reluctantly negotiate, mayors can. If citizens are defined by nations as spectators to their own destinies who think ideologically and divisively when they think at all, neighbors and citizens of towns and cities are active and engaged. They tend to think pragmatically and clearly, which is to say publicly and cooperatively; they think in the way theorists of democracy have always said they would in a well-constituted civil society that empowered them as members of a commons.

7
The Road to Global Governance

In May 2016, the tar-sands community of Fort McMurray, Alberta, paid a terrible price for the world's disinterment of carbon fossil fuels: a wildfire burned out the town, putting nearly 90,000 people in panicked flight, destroying 2,500 homes, and incinerating a half million acres of forest in a week. Record heat over northern Canada, combined with exceptionally dry conditions, helped turn an ordinary wildfire into a disaster. A few months later, as mean temperatures continued to rise and the retreat of glaciers accelerated, Delhi reached 123 degrees Fahrenheit for days on end, in what was the hottest year on record. Fire and melt combined to darken Greenland and Canada, making them less reflective of solar heat, compounding the problem of temperature rise. While it's impossible to link any one event to climate change, the size and frequency of major fires and the length of the wildfire season have all been increasing throughout the American West. Drought and hot weather—two effects of global warming—are the chief reasons.

Even as the climate threat grows, however, pessimism is beginning to wane. The response to the peril is finally becoming proportionate to its urgency. If we examine what cities and their networks are doing, often out of the public eye, it turns out to be worthy of attention and perhaps modest boasting. With the collaboration of people like Michael Bloomberg of New York (C40 cofounder and now U.N.

special climate envoy), Bill McKibben working with the Bernie Sanders campaign and the Democratic National Convention Platform Committee, Jeffrey Sachs of the Earth Institute (a U.N. adviser), Pope Francis, and such mayors (to name just a few) as Eduardo Paes of Rio, Park Won-soon of Seoul, and Anne Hidalgo of Paris, as well as the U.S. Conference of Mayors under the leadership of Tom Cochran, there is measurable progress on the urban climate agenda.

I have worried that the U.N.'s promotion of its Sustainable Development Goals (SDGs) would let member nations get away with promises and platitudes. But the Paris COP agreement and the accords coming from cities suggest, to the contrary, that real change can be animated by this broad declaration of principle. It is hard to imagine that China and the United States would have reached their bilateral accord (if only, in China's case, to peak carbon emissions by 2030) without the SDGs. That two nations accounting for 40 percent of the world's greenhouse gas emissions were able to come to an agreement clearly helped bring others to Paris—most significantly, China's sister "BRIC" nations Brazil, India, and Russia. Each had good reason, as a developing nation, to distrust appeals for climate action coming from the fully developed nations—the states responsible for most of the world's fossil fuel use and carbon emissions, which were nonetheless asking the Global South to foot much of the bill for warming caused by the North's development. Even in this disputatious arena, however, the recent Kigali amendment regulating HFCs (hydrofluorocarbons used in refrigeration), stages of development were taken into account in distributing responsibilities for compliance. (See below.)

Cities are hoping to give their laggard mother states, at long last lumbering into motion, the courage of their convictions. In April 2015, representatives of more than a hundred cities met in Seoul under the auspices of ICLEI—a long-standing association of cities committed to combating climate change—and adopted a new Seoul Declaration aimed at "building a world of local action for a sustainable urban future."[1] Previous documents, including the Compact of Mayors, the Earth Charter, and the criteria for participation in the C40 Cities attest to the robust urban engagement in sustainable action and resilience.[2]

The vigorous activity of cities does not start with but is dramatized by the story of COP 15 in Copenhagen. COP 15, with 184 nations in attendance, was supposed to be the last roundup for a worldwide agreement on climate change. It ended up, despite the hype, as still another failure. But urban municipalities were a different story. At the invitation of Copenhagen's mayor, Ritt Bjerregaard, a former European Union environmental minister, a couple hundred cities convened as a kind of rump urban parliament to consider what they might do as backstage players, however the nation-state negotiations turned out. With 80 percent of carbon emissions coming from metropolitan regions, it was clear to the mayors that cities could make a difference even if the nations managed no more than noble rhetoric. And they understood the need: the 90 percent of cities built on water (whether lakes, rivers, seas, or oceans) would be the first victims of climate-induced ocean rise, flooding, or drought.

A number of intercity associations were already developing urban decarbonization strategies that more recently became part of a new Urban Agenda (sponsored by Habitat III). The participants included older climate networks such as ICLEI, newer ones like the C40 Cities, and many general intercity associations with broader agendas where climate was still a concern. They included generic organizations such as CityNet (the Asian city network), City Protocol (the Barcelona-based web network partnered by Cisco Systems),

EuroCities, Metropolis, and UCLG as well as myriad national associations of cities similar to the U.S. Conference of Mayors and the National League of Cities.

The UCLG may be the most important global political body no one has ever heard of. Although it brings together thousands of cities around the world in an annual congress and ongoing cooperative projects and has been doing so since World War I, when it fused two nineteenth-century city networks, its name evokes blank stares among people who can easily identify the Concert of Nations, the League of Nations, and the United Nations. Yet there are dozens of such associations: in Europe, Energy Cities Europe, Strong Cities Network, the European Forum for Urban Security (EFUS), and the International Cities of Refuge Network (ICORN); in Asia, CityNet and the Mayors for Peace headquartered in Hiroshima; and globally, UCLG but also Metropolis, the U.N.-sponsored world organization of major metropolises, Cities for Mobility, CLAIR or the Council of Local Authorities for International Relations, and Delgosea, ICMA and INTA (global management and development organizations), the League of Historical Cities, Sister Cities International, and, under the United Nations, Habitat for Humanity, Cultural Cities, Green Cities, and many others.

No urban network, however, has had the impact of the growing handful devoted to combating climate change (Table 1). Among these, the best known and most effective have been ICLEI and the C40 Climate Leadership Group. The C40, founded in 2006 by a group of mayors led by London's Ken Livingstone, and including today not just forty but eighty cities, describes itself as a "network of large and engaged cities from around the world committed to implementing meaningful and sustainable climate-related actions locally that will help address climate change globally." Its leadership has

Table 1. Environmental City Networks

Organization Name	*Headquarters city, country*	*Membership*	*Year founded*	*Mission and Activities*
Alliance in the Alps www. alpenallianz. org	Mäder, Austria	300+ in 7 countries	1997	"An association of local authorities and regions from seven Alpine states. Its members, together with their citizens, strive to develop their alpine living environment in a sustainable way. 'Exchange–Address–Implement' is the main idea behind the Alliance's activities."
C40 Cities Climate Leadership Group live.c40cities. org/cities	New York (current chair city is Paris)	58 "global cities"	2005	The C40 is "committed to implementing meaningful and sustainable climate-related actions locally that will help address climate change globally." It engages a broad array of environmental and liveable city issues, including energy efficiency, emissions, waste reduction, bike infrastructure, public engagement and urban drainage. C40 Cities partners with the Clinton Climate Initiative (CCI).
Climate Alliance www. klimabuendnis. org	Frankfurt; Brussels	1,600 municipalities (in 18 countries)	1990	"European network of local authorities committed to the protection of the world's climate. The member cities and municipalities aim to reduce greenhouse gas emissions at their source. Their allies in this endeavour are the Indigenous Peoples of the rainforests in the Amazon Basin."

Covenant of Mayors (E.U.) www. eumayors.eu	Brussels	3,512 signatories (representing over 155 million citizens)	2008	The Covenant of Mayors is a "European movement involving local and regional authorities, voluntarily committing to increasing energy efficiency and use of renewable energy sources on their territories. . . . Covenant signatories aim to meet and exceed the European Union 20 percent CO_2 reduction objective by 2020." Signatories submit action plans and track their progress publicly on the website.
Energy Cities/ Energie-Cites www. energy-cities.eu	Besançon, France, and Brussels, Belgium	1,000 towns and cities in 30 countries	1990	"Energy Cities in the European Association of local authorities inventing their energy future." It helps cities "strengthen their role and skills in the field of sustainable energy," represents cities, sustainable energy at E.U. meetings, develop and promote initiatives through knowledge and experience exchange.
ICLEI–Local Governments for Sustainability www.iclei.org	Bonn, Germany	1,200 municipalities and associations from 70 countries	1990	"ICLEI is an international association of local governments as well as national and regional local government organizations who have made a commitment to sustainable development." ICLEI provides consulting, training, and platforms for information exchange to build capacity and support local initiatives to achieve sustainability objectives.

Organization Name	*Headquarters city, country*	*Membership*	*Year founded*	*Mission and Activities*
MedCities www.medcities.org	Barcelona	27 cities in 16 countries	1991	Medcities works "to strengthen the environmental and sustainable development management capability of local administration . . . and to identify the domains were a common activation could be the most useful mean to improve the regional environmental conditions." It aims "to reinforce the awareness of interdependence and common responsibility regarding the policies of urban environmental conservation in the Mediterranean basin."

involved Mayor David Miller of Toronto (Rob Ford's sane predecessor), former mayor Bloomberg of New York, and subsequent chairs, Mayor Eduardo Paes of Rio and currently Anne Hidalgo of Paris, as well as a nine-member steering committee.[3] Many in its leadership group have helped generate the policy ideas discussed in more detail later in this book.

While cities and their networks can achieve a great deal, much cross-border cooperation and informal governance grows out of voluntary actions undertaken by individual citizens and civic associations in response to common problems. Without active support from their citizens, city officials cannot be expected to make bold choices or take risks (including international collaboration) on behalf of the sometimes controversial policies that are required in dealing with climate change and sustainability, as well as with the adversaries of such efforts, who wield financial, media, and electoral power. The extraordinary civic support for sustainability policies that was expressed in the People's Climate March of September 2014, which brought more than 430,000 people into the streets of New York City and inspired parallel demonstrations in cities around the world, is a powerful example of what citizens can do. Climate movement politics culminated in the 2016 U.S. presidential campaign in Bernie Sanders's adoption of a strong climate agenda that found its way into the Democratic National Convention's policy platform. In late 2016, attention was focused on the Dakota Access Pipeline by the Standing Rock Sioux protest in North Dakota.

Bottom-up politics of the sort manifested in these popular movements is crucial in changing actual human behavior as well as governmental climate policy. Just as cities don't have to wait for states to achieve a measure of climate security or sustainability, civil society doesn't have to wait for city government. And citizens don't have to wait for civil society. The public square's many streets and the internet's virtual public square stand ready for those who would bypass traditional forms of political association. Citizens make up a global network in waiting and a potent and necessary foundation for global cooperation by cities.

8

Climate Justice: Making Sustainability and Resilience Complementary

It is important to recall that although they are often separated, arguments about sustainability cannot be segregated from arguments about justice and resilience. The origin of the Sustainable Development Goals in the earlier Millennium Development Goals demonstrate the closeness of this connection. The people most vulnerable to climate disasters that accompany global warming and extreme weather are those with the least mobility and those whose habitations are least resilient in the face of catastrophe. This puts at risk the poor, above all women and children, who are already endangered by their lowly status in traditional hierarchical (often patriarchal) societies. It was no accident that Ward Nine in New Orleans, among the poorest of the city's neighborhoods, suffered the worst casualties and was the slowest to rebuild after Hurricane Katrina. With neither automobiles nor even accessible public transportation, the residents' immobility in the face of the hurricane was in the most literal sense fatal. Rich and poor neighborhoods alike faced devastation, and wealth offered no immunity to death and destruction. But richer neighborhoods found a far quicker path back.

There is an unsettling connection between climate justice, sustainability, and the strategy of "resilience" as a response to climate change—the idea that we cannot prevent massive disruption from warming's effects, so our best course is to prepare to adapt. In the face of damage done or in the offing, cities are without question obligated to pursue policies that will enhance resilience. But fixing on resilience without linking it to sustainability can lead to injustice. First, the poor are the most likely to suffer the destructive consequences of warming, while cost-efficient resilient strategies are most likely to protect the privileged. The skewing of the economy can mean that "neutral" resilience strategies end up serving the wealthy first (as happens in education, housing, transportation, and other domains).[1] The poor are likely to benefit more from preventing climate change, which removes every person from danger, than from dealing with its consequences, which play out unevenly in an economically diverse population. A conscientious resilience approach aimed at equal protection would benefit everyone, but not every resilience approach is so conscientious. Given these inequalities, climate activist Tim DeChristopher argues, it is not "a coincidence that it's the groups from impoverished and oppressed areas or oppressed constituencies that are building the kind of [climate] movement we need."[2] Groups such as the Climate Justice Alliance, indigenous people's groups like Idle No More, and the Keystone XL pipeline resistance group in South Dakota as well as the Sioux in North Dakota are fighting what they call the "black snake" with one eye on carbon emissions and the other on injustice.

Climate justice can be a feature of a resilience strategy or not. The term is important, even fashionable, independent of one's attitude toward the justice question. The University of Nebraska Press has a new journal called *Resilience,* and Judith Rodin, the former president of the Rockefeller Foundation, has written a seminal book on strategies of resilience. She also launched an influential global program, 100 Resilient Cities, to help cities develop environmental resilience. The program funds the hiring of urban resilience officers and works with them to plan responses to climate catastrophes. By giving cities an incentive to act, its funding has a multiplier effect, yielding a significant impact in towns large and small.

Like all policy approaches that aspire to neutrality, however, 100 Resilient Cities cannot help but have a politics. Jill Abramson, the former executive editor of the *New York Times,* told Wake Forest University graduates in 2014 (reflecting on her own abrupt dismissal from the *Times*), "I cannot think of a better message for the Class of 2014 than that of resilience," even though she knows that women should not be expected to respond to unfair hiring and firing practices rooted at least in part in patriarchy solely by courting resilience. Our first option is always to challenge the untoward and intolerable consequences of attitudes and practices that can be changed up front.

In pioneering 100 Resilient Cities, the Rockefeller Foundation must take care that in its efforts to be inventive, adaptive, and entrepreneurial in *surviving* the coming climate crisis, cities do not overlook or push aside opportunities to prevent the crisis or preempt its causes. When the die is cast and the future is already fated to suffer the "chronic stress" and "acute shocks" resilience champions talk about, then their strategy becomes necessary for survival and the Rockefeller vision looks prophetic. Yet just as bomb shelters were once misused as strategies for coping resiliently with the threat of nuclear war, to which they actually helped accustom us, climate resilience can be misused to distract attention from sustainability and get us used to the "inevitability" of catastrophic global warming.

This is not to say that resilience strategies are unimportant: they will be crucial in confronting the ineluctable consequence of climate change. With scientists now estimating that the most progressive and forward-looking climate policies, if we somehow find the political will to adopt them, still will not prevent two or three degrees Celsius of temperature rise, resilience becomes indispensable. Even if we reduce carbon emissions to zero overnight, the carbon dioxide we have already put into the atmosphere will stay and stay, warming the globe even without any further emissions. The inevitable consequences of earlier CO_2 pollution render resilience strategies absolutely necessary.[3]

But from a political perspective, where resources are scarce and media information and civic education defective, and where cities feel pressed to choose between mitigation and prevention now or dealing resiliently with consequences later, some policy makers may opt for resilience at the expense of sustainability. Those prone to rationalize and greenwash their choices may divert attention from the fact that we ourselves are the cause today of the consequences for which we are seeking resilient responses tomorrow. Far from inevitable, the climate catastrophes to come are mostly man-made and (at least technically) preventable.

Resilience advocates are realists betting that the sustainability approach will fail. They read the skeptical account of democracy's capacity to cope, and they prepare for the worst. Who's to blame them? Anything less would be irresponsible. Yet resilience thinking can go too far. With better medicine, who has to worry about getting sick or waste time working at a fitness regimen? Preparing for the worst is buying insurance. In a world of human frailty nothing else makes sense, as long as we stay as keenly interested in achieving decarbonization as in addressing the consequences of failure.

That is what I mean when I say resilience can have a politics. As Timothy Noah has suggested, it may point to policies that are both "necessary and worthwhile," and the Rockefeller Foundation has proven that these policies can be indispensable to cities in particular, since they must confront changes and consequences for which they are not necessarily responsible.[4] Yet no resilience advocate who is serious about addressing climate change would place a premium on merely reacting to climate change at the expense of being proactive in prevention. The crucial need that Rodin insists moves us to address the consequences of "what goes wrong" must not stop us from trying to make it go right.

Even as the Rockefeller Foundation has used resilience to help cities prepare for the worst, Exxon, Shell Oil, and other carbon companies have been using it to make worse seem better, rationalizing their pernicious contributions to climate change by flying the resilience banner. In Rotterdam in 2014, Shell sponsored an international conference, a unique greenwashing event marked by (among other things) an "eco-marathon" for youngsters, a conference of experts including the president of the Rockefeller program Michael Berkowitz and me, to burnish the not always bright features of climate resilience.[5] Unlike the Rockefeller Foundation, Shell tends to fudge the crucial distinction between natural and man-made catastrophes. It depicts the climate crisis as an ineluctable feature of the human condition, alongside such unavoidable natural catastrophes as earthquakes and tsunamis. Shell's use of resilience is strictly diversionary, and it risks shifting responsibility from our profligacy today to what we hope will be the agility of our children and grandchildren in coping with the climate legacy we leave them. Resilience punts action on global warming forward into a future in which coping will be the only option, like a smoker counting on improved surgical procedures to excise his lung cancer in the future rather than quit smoking and prevent the disease today.

In the end we require both sustainability and resilience. Even Shell is finally acknowledging this. In May 2016, the company announced it was hedging against climate change by creating a new unit devoted exclusively to renewable and alternative energy. In doing so, it recognized that navigating climate survival by cultivating resilience alone was not enough. At the same time, in a world already on the way to a carbon dioxide level well over 400 parts per million (where 350 is the maximum "safe" level) and where the best efforts of COP 21 are being dismissed as insufficient, resilience strategies are absolutely required. But this world also demands an awareness of the ways in which a healthy resilience politics of mitigation and adaptation later can be perverted into an unhealthy and reactionary politics of inaction now. Why obsess hysterically over melting glaciers and vanishing coral reefs when tomorrow we can simply move to the mountains, eat less fish, and stop snorkeling? Better to manage the floodwaters to come, because it costs too many jobs or dollars or sleepless nights to curb carbon-induced sea rise today.

So as we cultivate resilience, with the help of thoughtful foundations and prudent city officials, in order to deal with the lowering of water tables and the rise of sea tides, more massive storms and more ruinous droughts, we also need to curb foolish practices such as growing thirsty crops like almonds in the deserts of California, or using precious water for fracking, or electrifying cars without reducing the number of coal-fired power plants, or recycling garbage without capturing the resulting methane.[6] We can farm smart with drip irrigation and tax carbon and subsidize alternative energy and limit emissions. We can push public utilities to accommodate solar rather than fight against it because their current systems can't handle the extra power.[7] And we can rescind those many invisible "Zombie Laws" that inadvertently stop cities from going green—aesthetic bans on clotheslines that force the use of electric driers, and on "granny flats" attached to suburban homes, expanding low-cost housing with a smaller energy footprint; or bans on strollers being carried on buses (forcing moms and dads to drive).[8] In short, in the cities where so much fuel is expended and so much housing needs to be insulated and so many people must be transported, the best resilience approach is to pursue sustainability now.

One cheer then for resilience: our last best hope should we fail to use our democratic institutions to forestall greenhouse gas emissions and limit climate change. Two cheers for democracy, when we use our common regulatory institutions to curb carbon production and use and diminish the need for resilience. And three cheers for sustainability, when we marshal cities to do everything possible today to make resilience a little less urgent tomorrow—when we bequeath to our children not only the knack for coping but a world in which coping is not the only option left.

9
The End of Sovereignty Redux: A Global Parliament of Mayors

If cities are to do what nation-states can't and won't, they will have to do more than they are doing, even though they are already doing a great deal. They will have to do it more collectively, and for that they will need more power than they have, more jurisdiction than they are permitted, and more resources than are returned to them by the "higher" jurisdictions whose coffers they fill but to whom they remain subsidiary. To get these things, they will have to advance their claim to govern globally in the language of rights: they will have to demand not merely the right of association but the right of collective self-governance. While cities already play a crucial role today in securing a democratic road to global sustainability, to be truly effective their authority and reach need to be amplified and globalized. It was for this reason that I ended *If Mayors Ruled the World* with a proposal for a Global Parliament of Mayors.

Surprisingly, given how hard it is to govern locally, mayors are also showing an appetite to govern globally, to govern together. Not out of ambition or pride but because they recognize that in an era of interdependence, local and global goods are pretty much the same thing. *Glocality* is proving that local government works more efficiently and productively when mayors cooperate globally, forging networks for common action. Yet glocality also reveals the irony of global challenges' requiring local solutions. Networked cities allow mayors to make the world their urban laboratory, and this is already happening. For many practical purposes, mayors already do govern the world and determine planetary outcomes. The urban world has for centuries been bound together by informal and formal ties that transcend national boundaries. Cities have always been defined by bridges rather than walls, connectivity not isolation.

Still, the myriad networks cities have constituted have lacked one feature: an explicitly political governance framework within which to realize their right to collective action; the deployment of common power on behalf of common purposes. As we have seen, existing networks, important as they are, often represent special issue silos, focused on urgent issues such as climate or security. Although they permit the sharing of best practices and the discussion of common strategies, and although they have yielded historic agreements like the Compact of Mayors, the Seoul Declaration, and the U.N. Habitat III "Urban Agenda" adopted in Quito at the end of 2016, they are not conceived as legislative or governance bodies, and they make no claim to exercise political power.

Long-standing associations like the UCLG do vital research and facilitate broad communication among cities. But the UCLG has not offered a platform for mayors themselves; it does the policy work and research that cities rely on for success. It and other urban networks have together built a sturdy foundation on which a governance body can finally be erected.

Such a body has been built. It met for the first time in September 2016 as the Global Parliament of Mayors (GPM), with more than seventy cities and two dozen urban networks participating, and will continue to convene not only physically but on an innovative new virtual platform that will permit mayors to meet on smart screens from their offices in hundreds of city halls around the world. Its initial steering committee, chaired by Mayor Jozias van Aartsen of The Hague, included Mannheim, Amman, Cape Town, Delhi (North), Oklahoma City, and Buenos Aires (the Tres de Febrero region). The GPM is no vague dream of idealists, but a hard product of urban pragmatists who refuse to stand by while the world goes to hell—or becomes hot enough to feel like it has. Participating cities include many with populations under a half million as well as metro-regions and megacities up to 30 million. They are port cities exposed to sea rise like Rotterdam, Tokyo, and Rio, and interior cities anxious about drought like Seoul, Oklahoma City, and Kampala; they are rich capital cities in the developed world like Paris, Atlanta, and Warsaw and they are burgeoning metropolises in the global south like Tijuana, Cape Town, and Dakkar. The GPM has little patience for national politics: Chinese and Cuban cities are no less welcome than Danish and Canadian ones. Towns with only fifty or a hundred thousand people forge urban six-packs that can be represented by a single spokesperson.

The development of urban networks over the past several decades, and of the Global Parliament of Mayors more recently, has cities poised to contribute significantly to a global campaign against climate change. The urgency of the challenge and the success at COP 21 have put nation-states on a more constructive road to change, but cities need not await good outcomes from national governments, especially as they come to represent reactionary populists averse to cooperation and skeptical of climate change. Cities are ready to engage across borders whether or not states are effective. The question is what exactly cities can do in concert that will reduce greenhouse gas emissions and induce decarbonization.

PART TWO
Making Democracy Work for Politics

• • •

"Anybody can set goals for 2050 or 2070—but we'll never reach them unless we start taking real action now."
—*Former New York mayor Michael Bloomberg*

"When national governments fail to act on crucial issues like climate change, cities have to do so."
—*Current New York mayor Bill de Blasio*

10

Common Principles and Urban Action

With or without the cooperation of the sovereign nations under which they govern and are governed, cities possess most of the tools needed to address and combat the challenges of global warming. The productivity and technology that accelerate warming can also be redirected to diminish it. A study by Bloomberg Philanthropies in 2014 concluded that "if cities act aggressively, they could reduce their annual carbon emissions by roughly four billion tons beyond what national governments are currently on track to do, in just 15 years."[1]

Having established the climate challenge facing cities and the political case for their taking action against it, in the second part of the book I want to examine some of the urban policies and sustainability strategies cities can actually use to reduce emissions and mitigate climate change, as well as prepare themselves to adapt to its consequences. There is no doubt that municipalities can do the job; the question is, can they do it together? And what exactly and how much can they can do in the absence of effective action by states?

There is no need to reinvent the wheel. Many of the policies for containing climate change have already been set out in important and well-known documents, such as those produced by U.N. Habitat and its latest Urban Agenda adopted in 2016 in Quito, the agreements pioneered by the Compact of Mayors and the C40 climate agreement, and by ICLEI in, for instance, its recent Seoul Declaration.

Others come from the work of research bodies like the Earth Institute at Columbia University, the Institute for Sustainable Development and International Relations (IDDRI) in France, and the Potsdam Institute for Climate Impact Research in Germany.

Nor do we need to rely on high ideals and altruism. As Michael Bloomberg has written, "for mayors, reducing carbon pollution is not an economic cost; it is a competitive necessity. . . . Dirty air is a major liability for a city's business environment."[2] The capacity and economic incentive are already present: what is needed to gain mayors' cooperation is a set of concrete proposals for turning these principles into a coherent action agenda.

The first approach to achieving common policies among diverse cities is to enunciate general principles and policies that apply across the urban landscape and may extend well beyond it. The Millennium Development Goals and their successor, the Sustainable Development Goals (SDGs), are an example. Goal 11 of the SDGs calls on signatories to "make cities and human settlements inclusive, safe, resilient and sustainable." That is an admirable sentiment, general enough to be universally acceptable among member cities. This principled approach may also be tailored specifically for urban settings. The C40 Climate Cities, for example, admonishes its members to reach a goal of "expanding climate action in cities" by reducing greenhouse gases and improving urban resilience over seven different sectors, including adaptation and water, energy, finance and economic development, measurement and planning, solid waste management, sustainable communities, and transportation.[3]

There are many other general principles of this kind: for example, Part II of the Earth Charter, which deals with a broad swath of environmental and ecological concerns, or the Sustainable Development Goals approved by the United Nations in 2015. These broad and encompassing statements are often rhetorically elegant, but they seldom secure much common action on climate change. The more agreeably bold and compelling a statement is, the vaguer and harder it is to apply. It may facilitate a thin consensus across communities and societies until people try to put into practice, at which point consensus vanishes.

Yet broad principles are just about always the platform on which major accords and agreements rest. The accord reached at the COP 21 meeting in Paris, for instance, goes back to the Earth Charter, an idea conceived in 1968 by members of the Club of Rome and reintroduced at the Rio Summit in 1994. It was finally approved in 2000 and has become a template for other declarations. The charter sets out its principles of ecological integrity in typically compelling normative language, here in four key articles on sustainability:

> Earth Charter Part II. ECOLOGICAL INTEGRITY
>
> 5. Protect and restore the integrity of Earth's ecological systems, with special concern for biological diversity and the natural processes that sustain life.
> 6. Prevent harm as the best method of environmental protection and, when knowledge is limited, apply a precautionary approach.
> 7. Adopt patterns of production, consumption, and reproduction that safeguard Earth's regenerative capacities, human rights, and community well-being.
> 8. Advance the study of ecological sustainability and promote the open exchange and wide application of the knowledge acquired.

These principles are easy to embrace precisely because they are broad and rhetorical, which is to say fuzzy about how they might actually work. There is a circular quality to the prescription: "protect the integrity of the earth . . . prevent harm . . . safeguard regenerative capacities . . . advance the study of sustainability"! Do these things and what have you accomplished? You will have protected the earth's integrity, prevented harm, safeguarded regenerative capacities, and advanced the study of sustainability. How precisely? Through which strategies? How do you address conflicts of interest? And how do you earn the trust of those who are disadvantaged by sustainability?

Such questions notwithstanding, prescriptive principle remains a useful starting place for securing some thin consensus among a range of players, from governmental bodies like states and cities to NGOs, to the private sector and even the energy industry. This is evident from the tone of the Earth Charter's impressive successors, such as the Millennium Development Goals established by the United Nations at the turn of the century, the Sustainable Development Goals approved in September 2015, and the Compact of Mayors established in 2014. The last is the world's largest cooperative effort among mayors and city officials to reduce greenhouse gas emissions, track progress, and prepare for the impacts of climate change.[4] It has established the Carbon Climate Registry as a central repository for climate and carbon data compiled from existing national, regional, and global city reporting platforms, mitigating the need for a city to report its climate data more than once.[5]

The original Millennium Development Goals included eight principles:

1. To eradicate extreme poverty and hunger;
2. To achieve universal primary education;
3. To promote gender equality and empower women;
4. To reduce child mortality;

5. To improve maternal health;
6. To combat HIV/AIDS, malaria, and other diseases;
7. To ensure environmental sustainability;
8. To develop a global partnership for development.

These are noble aspirations, unclouded by modesty or specificity. The Sustainable Development Goals—a far more elaborate set of seventeen principles and 125 subparagraphs—permit the proactive notion of sustainability to attach itself to nearly all domains of human behavior, but again without either specificity or relevance to actual policy practice. Some think the SDGs improve on the Millennial Development Goals in ways more audacious than useful. Here are the main principles stated as "goals" (I will spare the reader the subparagraphs):

Goal 1. End poverty in all its forms everywhere;
Goal 2. End hunger, achieve food security and improved nutrition, and promote sustainable agriculture;
Goal 3. Ensure healthy lives and promote well-being for all at all ages;
Goal 4. Ensure inclusive and equitable quality education and promote lifelong learning opportunities for all;
Goal 5. Achieve gender equality and empower all women and girls;
Goal 6. Ensure availability and sustainable management of water and sanitation for all;
Goal 7. Ensure access to affordable, reliable, sustainable, and modern energy for all;
Goal 8. Promote sustained, inclusive, and sustainable economic growth, full and productive employment, and decent work for all;

Goal 9. Build resilient infrastructure, promote inclusive and sustainable industrialization, and foster innovation;
Goal 10. Reduce inequality within and among countries;
Goal 11. Make cities and human settlements inclusive, safe, resilient, and sustainable;
Goal 12. Ensure sustainable consumption and production patterns;
Goal 13: Take urgent action to combat climate change and its impacts;
Goal 14. Conserve and sustainably use the oceans, seas, and marine resources for sustainable development;
Goal 15. Protect, restore, and promote sustainable use of terrestrial ecosystems, sustainably manage forests, combat desertification, and halt and reverse land degradation and halt biodiversity loss;
Goal 16. Promote peaceful and inclusive societies for sustainable development, provide access to justice for all, and build effective, accountable, and inclusive institutions at all levels;
Goal 17. Strengthen the means of implementation and revitalize the global partnership for sustainable development.

Goals 7, 8, and 9 address climate change directly, but in language so generic that embracing them requires little more than feigned good will. Seven additional goals (11 through 17) are intended to embed the principle in cities and communities, leading to somewhat greater specificity. Making cities "inclusive, safe, resilient, and sustainable" is worthy and persuasive, yet we have already noted that sustainability and resilience may require different approaches and different expenditures of resources. Nor will pursuing them automatically "reduce inequality," the key to Goal 10, which raises the question of climate justice.

Such issues leave critics skeptical. *The Economist* may be rather too cynical when it claims "the efforts of the SDG drafting committees are so sprawling and misconceived that the entire enterprise is being set up to fail."[6] But it appears that the SDGs are intentionally silent on issues where discord might arise, such as policy, measurement, and inspection. Nor do they try to provide common indices that might render different policy approaches commensurable.

This leaves them as little more than declarative prompts asking cities to do *something,* anything, soon, today, whenever. Moreover, for all their generality, they do not really respond to the underlying policy questions raised by the science of climate change. Neither Miami Beach nor New York would disagree with SDG 11's call to be "safe, resilient, and sustainable." But how? Safe for whom? Is it better to prohibit development on waterfronts imperiled by rising sea and surging storms? Erect artificial barriers—locks, dams, dikes—to keep the sea out? Or respond with natural strategies such as restoring wetlands and marshes or building dunes, even at the cost of dislocating some residents? Is it enough simply to raise the sidewalks on the seaward side of the street, hoping to protect the stores on the other side (as Miami Beach has done)?

Critics might also notice that SDG 11 stands in potential conflict not only with SDG 10, on reducing inequality, but with SDG 12, which calls for "sustainable consumption and development patterns." At least on the American model, "sustainable consumption and development" is an oxymoron when it comes to the deep sustainability the climate crisis demands. When compared to the broad-principle sustainability approach, resilience, which aims at dealing with the actual consequences of climate change if sustainable policy fails, offers a welcome practicality. Although resilience strategies can be more reactive than proactive, they do confront the climate threat without wishful thinking, giving mayors an important sense of relevance that "big principles" cannot impart.

What can be useful about charters and declarations of purpose is that they point back toward and dramatize the larger natural context of the Anthropocene, in which our species no longer acts as an integral part of nature but aims at dominion over it. Seen from this perspective, climate change is a critical but by no means the only challenge. It belongs to a set of planetary boundaries on which the earth's survival depends. In his climate encyclical, Pope Francis seems almost to have put the idea of planetary boundaries into theological language: Christians have misinterpreted scripture, he asserts, and "must forcefully reject the notion that our being created in God's image and given dominion over the earth justifies absolute domination over other creatures."[7]

The concept of planetary boundaries, introduced earlier, is an example of the power of such broad principles. These include climate change, ocean acidification, stratospheric ozone depletion, the nitrogen cycle, the phosphorus cycle, global freshwater use, change in land use, biodiversity loss, atmospheric aerosol loading, and chemical pollution.[8] They are interdependent and fragile, so that overstepping one can lead to weakening another.

Decarbonization is a strategy that can diminish these interdependent threats to a number of planetary boundaries at once. This gives even greater importance to the Decarbonization Project, a joint effort of fifteen countries working together to develop national pathways to economic development and energy production consistent with the two degree Celsius limit on global warming.[9] The three key standards by which decarbonization is measured are energy efficiency (greater output per unit of energy input); electrification of energy (reduce emissions of CO_2 per megawatt of electricity generated); and decarbonization of electricity (fuel sourcing shift away from fossil fuels).

Cities can develop comparable strategies aimed at decarbonization as well as sustainability and resilience to combat (and address the consequences of) climate change, but also to defend other planetary boundaries. Biodiversity is connected to chemical pollution, ocean acidification is linked to changes in land use, and global freshwater use is impacted by agriculture. Growing pecans that draw on huge water resources (300 gallons for one pecan!) in desert climates like Arizona for the enjoyment of consumers in California's water-short cities makes no environmental sense, either for farmers or for Los Angeles. Fracking both uses and contaminates water supplies, which makes it of concern to city dwellers well removed from regions where shale is being hydraulically fractured. Directly or indirectly, both practices contribute to or are affected by climate change.

Planetary boundaries are deeply and essentially interdependent. As preamble to an urban action agenda, generic strategies responding to the broader challenges of planetary boundaries are important guides to spirit and intention, and entail greater specificity with respect to climate action for both nations and cities than either the SDGs or various urban declarations. That is one of the virtues of Jeffrey Sachs's approach. Yet translating such broad-stroke concerns into actionable urban policies on which there can be agreement across the diversity of global cities remains challenging. For here is the conundrum: when standards are encompassing enough to be relevant in *every* urban context, they are relatively void of content in *any particular* context.

11
The Politics of Commensurability and the Challenge of Trust

The second approach to creating standards for urban action on climate change is to embrace a diverse menu of options open to all cities, and either seek a common policy that fits all or render different policies fair and useful and thus trustworthy through an index of commensurability. This strategy recognizes urban diversity while offering the hope of flexible policies that engender trust and further cooperation. If we can get New York, Copenhagen, and Seoul to agree on WtE (Waste to Energy) or Bike Share programs, that is useful. But not all cities are likely to agree. Flat, dense Manhattan is a better bet for bikes than hilly San Francisco or sprawling L.A. And if New York decides to paint its tar roofs white while Copenhagen focuses on LEED insulation and Seoul does river development (both elaborated below), they obviously need to find some way to align and compare the three strategies in terms of a common standard of sustainability that all three municipalities can accept.

Cooperation, while necessary, is by no means easy. The profound differences among cities make it hard for them to move from broad sustainability principles to specific policies. Once cities have reached consensus on broad general principles, the trick is to develop an agenda with enough flexibility and variety of options to allow for urban diversity. And there is a third and more daunting necessity: the obligation to fairness. Cities, no less than nations, need to trust one another. To cooperate, they must believe that their distinctive approaches to climate change are commensurable: that their efforts and expense are measured against some fair and transparent standard that all can accept. There must be some agreed-upon trade-off between, say, a rapid transit light-rail system like Bogotá's and a waste to energy focus like Oslo's. Such indices allow a meaningful ranking of different cities' contributions to curbing climate change, even if their policy approaches are very diverse. Without the assurance of fairness and accountability, collaboration will fail.

No one policy or strategy can work for every city—although some, such as above-ground mining or car-free pedestrian zones, have more general applicability than others—and there are many ways to account for and evaluate the differences. Comparisons are likely to be subjective; classification schemes are often at odds. In an assessment of the Siemens Index discussed below, for example, the environmental strategist F. Kaid Benfield commented that "any attempt to measure, score, or rank places with respect to almost anything will be incomplete at best and can be wildly misleading at worst." Although Benfield has himself designed a neighborhood development LEED measure (Leadership in Energy and Environmental Design), he admits that "rating systems tend to assign numerical grades to things that are partially or entirely subjective" and that "measurements based on qualitative data are complicated." Fortunately, the weighting need not be perfect; it need only create trust by demonstrating fairness across different actors and activities. Still, persuading cities to acknowledge the comparisons may require the very trust commensurability is intended to produce.[1]

So we must enter the world of comparing cities around climate action with a good deal of skepticism. Intercity comparisons aimed at classifying common features are challenging in almost any domain. A recent article in the Royal Society journal *Interface,* for example, argued that in comparing urban design on the basis of mathematical models, one might posit four distinct categories of urban design: medium-sized rectangular blocks like Buenos Aires (the only city named in this category); small blocks with a diversity of shapes, like Athens; larger, more balanced blocks with a diversity of shapes, like New Orleans; and a mosaic of small squares, like Mogadishu. Yet while "you can think of this classification being for cities what zoology is for animals or botany for plants, a way to sort some objects according to their similarity," we can easily imagine scores of alternative classifications, not one of which is without controversy.[2]

We can, for example, be much more generic and classify cities according to whether they are on water or landlocked, big (or mega-big) or medium or small, sprawling or dense (horizontal or vertical), wealthy or poor, old or new, centrally integrated or districted into neighborhoods, planned or a product of "organic" development. There is no right or wrong; it depends what you are looking for. But how a city is defined by such categories can shape how leaders view the content and scope of policy options.

Even superficially similar strategies may be incommensurable in practice. A bike-sharing program in, say, the mega-region of Beijing (projected to be growing toward 130 million people!) may be so different, in terms of its logistical, administrative, financial, and cultural burden, from a bike-sharing program in Louisville (population just under 250,000) that there is hardly a fair basis for comparison. Yet to persuade two such cities to cooperate on a bike-share or any other climate strategy, we must identify some common metrics against which their diverse policies or distinctive circumstances can be measured. There are so many different ways cities can view their sustainability policy options: through the lens of the three P's of sustainability (people, profit, planet); or through the optics of the three so-called ESG domains, the economic, the social, and governance. Or through mixing and matching by choosing a key frame from among these categories. This book leans heavily on people (as citizens, that is, democracy), planet (long-term world sustainability), and governance (the politics of climate change). Another book focused on the carbon market might have highlighted the economics of sustainability, while a study of popular movements and climate would have examined civil society.

So this is the dilemma: cities committed to developing common urban policies on sustainability and climate change must make collective decisions. Yet to do so they must reconcile the reality of urban diversity with the need for common measures by which radically different strategies and policies can be compared and evaluated across regions, states, and continents. The challenge is concrete and altogether real. You cannot take cities with ample hydroelectric resources (Zurich, for example) or nuclear energy plants (such as Lyons, France, where nuclear plants generate some 70 percent of energy) and measure their "progress" on decarbonization in the same way it is measured in Calgary or Warsaw, where the primary energy generators are heavy oil and coal. By the same logic, cities with emerging economies like Chongqing or Sao Paulo cannot be held to the same standards as postindustrial cities such as Osaka or Atlanta, whose "development phase" is behind them.

Nations recognized the importance of stages of development in the Kigali amendment on hydrofluorocarbon refrigeration and air-conditioning gases (see Part I above) by establishing three classes of treaty signatories: developed nations, committed to immediate reductions, a middle class of BRICs (minus India) and others on a slower timetable commensurate with their need as developing economies for inexpensive refrigerants; and a third class of nations like India, Pakistan, and African countries both less developed and in hotter zones where air-conditioning is indispensable. Tellingly, African nations voluntarily abjured their assigned status in the third class and opted into the second, signaling their understanding that the climate crisis required action now not later, above all in the hot zones where warming's effect would be most potent.

Cities patently need comparable approaches to their differences. Greening a port may work for Rotterdam or Los Angeles, but how to compare it with insulating the extensive traditional building infrastructure of New York? Sprawling horizontal cities like Beijing, Mexico City, or Phoenix depend more on automobiles than dense cities served by efficient undergrounds such as New York or Paris, but the horizontal cities also respond better to new transportation options like the cable cars or bus-based above-ground rapid transit systems pioneered in Medellín and Bogotá. How then to establish criteria without seeming to favor some municipalities but not others? And how to assure that the science of what is possible does not occlude the politics of what is doable, that is to say political will?

Important as it is, the scientific depiction of technical responses to crises can have a wonkish and cold-blooded feel that inspires agreement among the well-informed but incites little action among politicians and ordinary citizens worried about economic costs and political liabilities. Commensurability, even when secured, need not produce trust, and trust can arise out of many causes besides commensurability. Environmental reasoning will not by itself create a platform for urban action.

Many a mayor has endorsed in theory the ideal of a walkable city with bike lanes and pedestrian landscapes, only to find disgruntled shop owners and angry cab drivers in opposition, threatening to exact high political costs if their objections are not heeded. It is easy for populists to persuade agitated and resentful followers that climate change is some kind of hoax being perpetrated by elites to advance their political careers at the expense of working-class jobs. Although about 70 percent of Americans now accept that global warming is happening, many fewer regard it as a salient issue. In 2016, a George Mason poll reported that climate change ranked near the very bottom, 21st in importance of 23 issues surveyed, among Donald Trump's supporters. (As noted earlier, Trump had called global warming a Chinese hoax.) At one point in his primary campaign, Trump said "the concept of global warming was created by and for the Chinese in order to make U.S. manufacturing non-competitive." He later disavowed the comment as a "joke" but still insisted he would "renegotiate" the COP 21 "deal," which he is now in a position to try to do.

These are unfortunately not simply instances of a single candidate's disregard for facts. They are part of a long-standing politics of denial that has crippled the response of the United States to climate change over decades. Republican Party climate know-nothingism goes back at least to 1992, when George H. W. Bush described that year's Democratic vice-presidential candidate, Al Gore, as someone "so far out on the environmental extreme, we'll be up to our necks in owls." And in the 2016 Republican primary campaign, it was not just Trump but also his running mate, Michael Pence, who stood as an avid climate denier; someone who had also rejected the link between smoking and cancer. One of Trump's most prominent opponents, Senator Ted Cruz of Texas, told a CNBC reporter that climate change was "the perfect pseudoscientific theory for a power-hungry politician."[3] By then, this had apparently become the required stance for any Republican hoping to be nominated for president. Denial was the hard tactic on climate, but a more powerful soft tactic was also gaining traction: doubt. As the book *Merchants of Doubt* by Naomi Oreskes and Erik Conway showed, spreading uncertainty under the guise of scientific skepticism can be much more effective than denying the conclusions reached by science. With Donald Trump, the view is now firmly ensconced in the White House.

In this setting, the futility of talking about the ample new technologies in place to inaugurate a campaign to curb climate change proved once again that it is politics, not science or technology, that stands in the way. Not even the welcome moral passion of Pope Francis's *Laudato Si'* has overcome green politics gridlock, although the pope certainly moved beyond the wonkish brain to the human heart, much as Rachel Carson and Bill McKibben did in earlier epochs. The political importance of the moral voice has been evident in the influential role Pope Francis has played in the urban battle against climate change, and also reflected in McKibben's recent role in the Democratic Party's 2016 presidential campaign.[4]

No document declaring general principles has generated more ethical heat or incited more political controversy than *Laudato Si'*, a tribute to the fact that it makes bold moral claims yet manages to couch them in quite concrete social and economic terms—terms that do not shy away from shaming climate deniers and corporate malefactors. "The warming caused by huge consumption on the part of some rich countries has repercussions on the poorest areas of the world," the encyclical reads, "especially Africa, where a rise in temperature, together with drought, has proved devastating for farming." More surprisingly, Pope Francis chooses to name names, identifying not only what must be done but to whom: who must be opposed and what must be overcome. "By itself," Francis insists, "the market cannot guarantee integral human development and social inclusion." Climate change cannot and will not be challenged until we "break out of the logic of violence, exploitation, and selfishness."[5] Yes, we must prevent diminishing the planet's biodiversity; no, ocean acidification is not of tangential importance. On July 21, 2015, at his Vatican City convening, Pope Francis actually framed the subject as "modern slavery and climate change," making the obvious but rarely mentioned connection between climate change and justice and inequality, a theme central to *Laudato Si'* and one to which cities have been more attuned than national governments.

Moral language actually plays little part in the anodyne pledges of most climate declarations and broad setting of "goals." The encyclical is not only more ethical but also more political than most declarations, arguing that action depends upon an awareness of the link between climate and justice. In this it has more in common with Naomi Klein's radical polemic *This Changes Everything* than with Jeffrey Sachs's careful study *The Age of Sustainable Development.* Klein too sees climate change as intimately bound with social justice. Her vision of the future "goes beyond just surviving or enduring climate change, beyond 'mitigating' and 'adapting' to it in the grim language of the United Nations. It is a vision in which we collectively use the crisis to leap somewhere that seems, frankly, better than were we are right now."[6] Scientific understanding and agreed-upon goals are then necessary but insufficient goads to action. Political will, even when incited, must be matched by political institutions and common behaviors capable of instigating action. As Pope Francis writes, there is "an urgent need of a true world political authority." That authority does not currently exist.[7]

Since Pope Francis wrote those words, a first iteration of such a world political authority has been realized: the Global Parliament of Mayors (GPM). The GPM, first convened in The Hague, September 10–12, 2016, and scheduled to meet again in the Global South in 2017 as well as on an intermittent virtual basis on a specially designed digital platform, is up and running, having already participated in Habitat III in Quito (October 2016) and in other international forums. Only a global authority, whether the GPM or others such as ICLEI or the C40 Climate Cities, that deploys common power and expresses a collective urban voice beyond what the established urban networks are able to do, or than nation-based global organizations have done, can forge a common urban climate agenda. This means an agenda that is both specific and particular to the differential needs of cities but at the same time capable of achieving consensus. The parliament is a new stage in the development of urban networks, but it has hardly been constructed ex nihilo. There is already a robust and active complex of urban networks dedicated to global climate action among cooperating cities. Without them, no global platform for cities would be possible.

12

City Sovereignty and the Need for Urban Networks

The Compact of Mayors launched at the United Nations Climate Summit in 2014 engaged "the world's largest coalition of city leaders addressing climate change by pledging to reduce their greenhouse gas emissions, tracking their progress, and preparing for the impacts of climate change." Eighty-four cities, with more than a quarter of a billion in total population, initially signed on, undergirded by such key networks as ICLEI, the C40 Climate Cities, the UCLG, and the Rockefeller Foundation's newly launched 100 Resilient Cities program. These and other networks fostered support for the global climate agreement reached at the COP 21 meeting at the end of 2015.[1] The next year, the European Union's Committee on the Regions added an Urban Agenda and a "Pact of Amsterdam" to the emerging consensus on the core role cities must now play in grappling with climate change, refugees, pandemic disease, terrorism, and global inequality.

These urban networks, though sometimes diverted by the politics of turf, are committed to a common mission—to get member cities to meet specific climate goals and environmental standards as a starting point for cooperation and common action—which requires their concerted effort if it is to be realized. But even with the most robust cooperation, cities in these networks have lacked sufficient jurisdictional autonomy and resources to be able to forge common policies across national borders. By themselves, cities and networks can only do so much. They are far too often prevented by national governmental policies from engaging in cross-border common action. Richard Schragger has argued that at least in the United States, "the vertical separation of powers . . . does not promote city power; it limits it," and that "if one favors decentralization, federalism as practiced in the U.S. is not the way to get it."[2] However, while cities have yet to flex their municipal muscles to secure the political mandate that will afford them the transnational authority and legitimacy to act collectively in place of their inert but still sovereign mother nations, their right to do so is increasingly evident.

Unfortunately, national dysfunction has not necessarily led to national modesty. As Michael Bloomberg has said with respect to sourcing energy, countries must "empower their cities to achieve [environmental] goals by freeing them to regulate their own power supplies." Cities need to be freed to take common action where states fail to act, as they so often do. In Bloomberg's derisive formulation, "federal legislators, as usual, are way behind the curve—laughably setting goals for some far off time when they'll all be dead and can't be held accountable!" Bloomberg's successor, Mayor Bill de Blasio, goes even further: "when national governments fail to act on crucial issues like climate change, cities have to do so." But to do so, they need to act in concert through a global political authority—which is the whole point of the Global Parliament of Mayors. In a task as "monumental as it is essential," Bloomberg adds, the new "Metropolitan Generation" will be required to "build modern cities for a new urban civilization." This is a challenge, like the challenge of climate change, that cities can meet only by working together.[3]

With the reputation of cities as agents of sustainability rising, and with their networks approaching the limits of their current legitimacy under traditional definitions of national sovereignty and subsidiarity, advocates of urban climate action continue to seek new forums where they can have an effect. In the summer of 2015, for example, Mayor Anne Hidalgo of Paris (along with the ubiquitous Michael Bloomberg) convened a mayors conference on climate in Lyons, France. The Lyons meeting anticipated the December COP 21 meeting in Paris, and as the two hosts made clear, it was inspired by the recognition that "the bold actions taken not only by local leaders but also by all the range of non-state actors to reduce greenhouse gases place them at the forefront of the fight against climate change." Their planning resulted in a "Climate Summit for Local Leaders" held in parallel to the COP 21 governmental summit.[4]

Both Hidalgo and Bloomberg speak from a foundation of action by their home cities. New York, for example, has realized a 19 percent reduction in carbon emissions in just six years, and Mayor de Blasio has pledged still more, including a projected multi-borough light rail line.[5] Paris meanwhile has set the audacious goal of a 40 percent reduction by 2030, and has also borrowed New York's idea of painting roofs white to enhance reflectivity and diminish heat absorption in buildings. Compare these measurable reductions by cities with the bilateral agreement between the U.S. and China, in which the United States ambitiously committed to do—well, what it was already doing anyway to reduce emissions. China, meanwhile, not quite so bold, did not pledge to decrease emissions, only to stop increasing them around 2030. That this exceedingly modest agreement was met with universal acclaim was a tribute to how little is expected from nations today, in this case with one in bondage to coal and the other to carbon companies and their misleading appeal to natural gas (and hence shale fracking) as "transitional strategies."

At the same time that Bloomberg and Hidalgo were hosting city representatives in Lyons, Pope Francis was convening seventy-five mayors, including many from ICLEI and the C40, at Vatican City to explore the connections between climate change and justice he had introduced in *Laudato Si'*. Among those attending was Mayor de Blasio, who addressed the gathering with fresh energy, suggesting that New York's commitment to city action on climate was not limited to former mayor Bloomberg but a continuing priority. "We, local leaders have many tools," he said, "more than we in fact realize, and we must use them boldly—even as our national governments hesitate." He then announced the next big step: "we are committing to 40 by 30 [40 percent reduction in emissions by 2030], on our way to 80 by 50 [80 percent by 2050]."[6]

It was thus not a complete surprise that when the United Nations enticed nearly two hundred anxious and reluctant nation-states to Paris at the end of 2015 for still another COP meeting, cities pointed the way to a climate agreement of some specificity. The reality of climate change was no longer a threat from gloomy scientists but a visible fact evident in melting glaciers, extreme storms, and escalating tidal surges, and this fact weighed on prime ministers and presidents even as mayors guided them like seeing-eye dogs toward the light. Mayors knew well enough that the diffident steps promised by the COP 21 treaty signaled not the end of their work but a new beginning; that if they did not move aggressively not only to implement the very modest aims of COP 21 but to accelerate the pace of mitigation and decarbonization, their citizens would suffer dire consequences.

Yet here we come full circle: cities, however willing, still find themselves impeded at every turn in taking on the global challenges of climate change by the still belligerent claims of nation-state sovereignty being made by national leaders who have in fact defaulted on their sovereign obligations. Populists and nationalists appeal to sovereignty lost, and supporters of Brexit and stubborn advocates of Trumpism are certain they can "restore" that lost authority of independent nation-states. Trump himself suggested in the last days of his campaign that Hillary Clinton was meeting "in secret with international banks to plot the destruction of U.S. sovereignty."

The decline of national sovereignty, however, is not an outcome of sinister plots being fomented by pushy globalists or disloyal cosmopolitans, it is the irreversible consequence of interdependence. It is the result of the permeability of national borders and jurisdiction in a world defined by transnational forces, from climate change and trade to terrorism and markets. The world's governments can no more secure their citizens against the devastation of warming than the American government can protect its people from firms moving jobs and capital abroad. It is ironic then that, attenuated as it has become, national sovereignty is still regarded as the presumed barrier standing in the way of municipal action across borders (and often within them as well).

The constitutional subsidiarity of cities to national governments that have lost their capacity to protect and sustain cities remains a persistent hindrance to global urban action on climate change. Although cities act cooperatively across borders with far greater facility than nation-states, their will to cooperative action collides with the supposed "independence" of states insisting on a de jure sovereignty that no longer obtains in reality. England chooses to exit Europe out of nostalgia for its former imperial sovereignty, which has ceased to exist for reasons that have little to do with the European Union. London and Leicester will complain bitterly, knowing interdependence is the lot of an urbanized and globalized world, while nationalists, nativists, and reactionary populists will preach the foolish gospel of borders and walls, although they mean nothing to pandemic diseases or drugs or refugees or terrorism or sea rise or the Internet. The sovereigns cannot govern, but they can still ensure a paucity of municipal resources and jurisdictional competence that makes it impossible for cities to act aggressively and collectively.

Things are changing, however, precisely because such threats as climate change, refugees, and terrorism demand that cities take action no prudent national government will want to inhibit. The European Union's Committee on the Regions (CoR) has traditionally recognized that regions and municipalities, though subsidiary to the union, are crucial instruments of policy. With the recent erosion of the European Union's authority in the face of populist and nativist anger at mass immigration by refugees, the CoR has moved to offer cities greater autonomy. In an effort to forge an Urban Agenda for Europe (sanctioned by the Pact of Amsterdam in spring 2016), the CoR acknowledged that it needed a more balanced and reciprocal relationship with its cities. What began, for example, as an effort to establish "an EU Urban Agenda" as a top-down gift to its cities became the more bottom-up "Urban Agenda for the EU." The seemingly subtle name change actually reflected a profound movement in the attitude of Brussels bureaucrats, as well as a willingness by cities to assert greater authority—not soon enough, however, to preempt Brexit.

In the United States, momentum has also been going to the cities, although this is most apparent in the visceral reaction of state and national governments against entrepreneurial city action. In the last years of his second term, President Obama worked with and through cities to secure what he could of his national agenda on equality, health, and other key domestic policies. And cities were looking to take matters into their own hands on climate. Fracking and drilling bans were passed in New York state, Pittsburgh, Denver, and smaller cities in Texas and California. And following the lead of several Scandinavian cities, including Oslo, Copenhagen, Uppsala, and Stockholm, American cities are being pushed by citizen-based climate organizations such as 350.org to consider divestment from carbon polluters. The potency of these actions, even as a threat, can be seen in the strength of the political and legal reactions to them. Courts are moving aggressively against the local drilling bans. Cities on the Rocky Mountain Front Range that have tried to restrict hydraulic fracking, for example, have been faced with a Colorado State Supreme Court decision overruling the ban imposed by Denver and enjoining Fort Collins and Longmont from enacting their own prohibitions. The Texas State Legislature got out in front of its cities in 2015, passing a law preempting all local control on a broad swath of drilling activities. Nonetheless, cities continue to fight: climate advocacy groups in California put a fracking ban on the state referendum ballot, and despite the animosity of courts and politicians fed by fossil fuel companies, cities persist in believing they have an obligation to protect their citizens.

In New York City, for example, after a combative two-year campaign, the City Council, following the lead of California and Washington, D.C., managed to pass a bill imposing a small fee on disposable plastic shopping bags, to discourage their widespread use and encourage environmentally safer alternatives. (Plastic bags are made from petroleum, and they foul waterways and pose a danger to birds and fish.) There were exemptions for the poor to prevent the fee from becoming a regressive tax. Not surprisingly, within a few weeks of the City Council's action, it was put on hold by the state legislature in Albany. Why? As the *New York Times* editorial page commented, because "it can." In many different ways, the mayor of New York must bend to the will of the governor and the state legislature. The city, wrote the *Times,* "has no power compared with the muscle Albany can flex." The "will of a sovereign city" meant little to state politicians under the influence of "plastic bag-maker lobbyists."[7]

Even as the concept of "city sovereignty" grows sufficiently widespread to be acknowledged by New York's (also America's) leading newspaper, its compass is being restricted by provinces, states, and national governments, and not just in the domain of climate change. Urban shooting deaths, especially among young people of color and in urban gangs, have reached epidemic levels in Chicago and Washington, but cities have been enjoined from protecting their citizens by federal courts enforcing a new and historically doubtful interpretation of the Second Amendment. A federal court ruled in May 2016 that people who apply for a license to carry a gun in Washington, D.C., must no longer be asked by the police to provide a "good reason" why they need one.[8] The month this ruling was issued, the even more violence-plagued city of Chicago was setting records: there had already been 1,500 shootings in the city that calendar year. In 2015 in Chicago, 2,998 people were shot, almost all of them young men of color, many in gangs. City officials around the country were in effect scolded for thinking they had a right to try to protect their citizens and residents from the severe threat to urban stability and safety represented by the gun lobby, hiding behind a Second Amendment conceived a couple of hundred years ago in an age of muskets and local militias.

In the domain of discrimination and human rights, American cities are also seeing their power to act restricted by state laws. In March 2016, the North Carolina Assembly passed a bill, purportedly to deal with a controversy over the use of bathrooms by transgender people, that denied cities throughout the state any right to define classes of people in order to defend them against discrimination. This prerogative was reserved to the state, which used it to narrow its definition of who could be protected. The bill, known as HB2, not only defied the federal government's intervention on behalf of LGBT rights but denied cities any right whatsoever to act independently in the domain of discrimination. HB2 not only nullified municipal statutes protecting LGBT rights but effectively denied cities the right even to set a minimum wage different from the state's. It becomes a question for cities whether they are obligated to obey international laws and the mandate of universal human rights, and if so whether they have a commensurable obligation to disobey national laws contrary to international law and human rights.[9] This was at the root of Mayor de Blasio's effort to defend urban rights against potential abuses by a Trump White House. It seems clear to me that as a function of their own shared sovereignty and their commitment to protect the rights of their citizens they are so obliged, at least when the urban view of rights is in accord with universal rights as reflected by international law and contradicted by parochial state and national notions denying such rights.[10]

Although sovereignty may no longer define the actual capacity of nations to guarantee the lives, liberties, and property of their citizens (which is, after all, the condition on which nations' right to govern is founded), it remains the legal framework for the denial of urban rights in America by the national government.[11] Indeed, both the national government and the states (which the national government may sometimes be in conflict with, as in North Carolina) insist on denying cities the right of action on issues critical to their citizens, whether or not the cities being preempted are better positioned to take action. New York City can forge a public-private partnership to plant a million trees, and do it quickly, as Mayor

Bloomberg did in 2007; or mandate repainting all of those black tar tenement roofs around the five boroughs white to make them more energy efficient, as New York has done, and Paris is now doing. And it can do these things almost overnight, rather than through a twenty-year Conference of the Parties approach. Yet this promise of urban power collectively deployed is still pitted against the power of sovereign states to nullify municipal actions and block collective urban action.

It is apparent that cities need to put muscle on the bones of their hypothetical rights, above all when it comes to an existential threat like climate change. Given the history of sovereignty's noisy national claims and the ongoing pressure applied to cities by higher jurisdictions, mayors—cautious and conservative by nature—are understandably vulnerable to intimidation. Lacking the resources to govern their own cities (many of those resources having been turned over as tax revenue to the national government), and worrying that responsible action on sustainability might transgress traditional sovereignty and bring angry national (or provincial and state) governments down on them, they proceed with extreme caution.

Cities' place in the struggle for sustainability and resilience nevertheless is undeniable. In the language of Hidalgo and Bloomberg:

1. Local governments are the first responders to crises caused by climate disruption. They confront heat waves, floods, and hurricanes. To react effectively and prevent their citizens from the consequences of these "natural" catastrophes, local governments must be given new tools to reduce greenhouse gas emissions and adapt to climate change, and protect natural environment and biodiversity.

2. In that regard, greater fiscal autonomy is a key for success. Lima, for instance, was able to finance a major modernization of its mass transit system after it earned a credit rating that allowed it to borrow money in the capital markets.
3. The more nations empower their cities, the bolder they will be. Cities with authority over building energy standards, for instance, deliver three times the results of cities that lack such power.[12]

The question remains how to assure that cities have this authority and are not obliged to wait passively for national governments to cede it to them. I have made the case for their right to act, but they need the resources, jurisdiction, and political will to do so. It has never been clear to me why the cities that generate the lion's share of tax revenues to provincial and national governments somehow become beggars when it comes to funding, often estimated to receive back less than half of what they pay in taxes. Without resources, their obligations quickly become "unfunded mandates," where they do a national government's bidding without the national government's resources. In economic terms, they are fighting above their weight while being funded far below their needs. The governor of Ohio can instruct his state's great cities (and lesser ones too) that state law requires them to replace their aging sewer systems, but give them no means to pay for this major infrastructure project. They are told to pay for it themselves with what is left after their assets have been taxed by the state.

Higher jurisdictions argue that cities need not be funded because they lack jurisdiction and authority in the domains where they are acting, even if the state is requiring them to act! German cities do not even control their local police, but they are burdened with all kinds of security and public order issues that are costly to address. American cities must deal with the proliferation of guns and pay for the consequences without the capacity to regulate firearms. Houston is subject to extreme flooding, but the Texas legislature has barred the city from confronting those who cause global warming; instead it leaves the metropolis to plead for funding as a supplicant when disaster strikes. What are cities to do? They have rights and obligations, as well as capacities. What they lack is permission from a sovereign state that isn't. It is the main political argument of this book that cities must act in common. What they cannot do individually, they must do together. In concert they can secure the abstract rights that, for one city at a time, are inaccessible in practice. Hence the call for collective muscle and a common political voice.

The missing link has been a viable global governance body. Fear of awakening aging states, those sleeping giants that prize the sovereignty on which they have already defaulted, has stopped cities from acting. The courage they need can only come from establishing a governing organization with the collective muscle not only to take on the global challenges of climate change, inequality, and terrorism, but to take on the intransigent nations unwilling to let cities rescue them.

Summarizing our argument up to this point, then, the task is to fashion and enact common urban climate policies that move beyond the generalities of declarative principle without stumbling on the particular features and parochialisms of diverse member cities. And then to establish legitimate global governing institutions that give cities a cloak of right to legitimize the power of common action they have already begun to exercise. The next chapter describes the areas in which cities are already developing responses to climate change and strategies aimed at both sustainability and resilience. We will then examine how a variety of cities are implementing specific policies aimed at curbing emissions, encouraging decarbonization, and slowing climate change. In the two final chapters I will take up, first, the tough issue of how to secure commensurability among the diverse and particular measures cities can take one by one, even when global consensus is absent. Second, I will provide a rationale for a new governing authority, without which the ample possibilities of individual collective urban action on climate will not be realized.

13

A Practical Climate Action Agenda

There is an extraordinary range of options available to cities committed to acting vigorously to combat climate change, even if they act alone. Many are well known, including for example, bike-share, recycling, and insulation programs (the "LEED" standards); others apply to regional and national as well as municipal governments, such as a carbon tax and divesting public funds from fossil fuel companies. But many remarkable, even astonishing efforts particular to individual cities are less familiar. The list below, though by no means complete, suggests what mayors can do when they put their minds and resources to work on it.

Fifteen Policy Options for Cities Pursuing Sustainability

1. *Divestment* of public funds from carbon energy companies. Cities have impressive municipal savings and investment funds, including public pension funds, public university and cultural institution accounts, and other funds that they have the discretion to invest in accord with their goals and values as well as simple profitability.[1] City pension funds in the United States alone are worth more than three trillion dollars. A fossil fuel divestment strategy that uses the market rather than government coercion can profoundly affect the profits of big oil and influence both corporate executives and shareholders. The

State of New York has pursued a fossil fuel divestment strategy, as has Melbourne, Australia. In Scandinavia, the world model region for this strategy, Malmo, Copenhagen, and Uppsala have all committed to divestment; in 2016 they were joined by Stockholm, where a citizen-led Fossil Free Stockholm movement helped lead the city to divest more than thirty million Kroner from oil, coal, and gas companies.

2. *Investment* strategies offer a parallel free-market option to divestment. Without requiring punitive laws, they enable cities to transform their market leverage into a tool of sustainability and decarbonization by encouraging renewable energy, infrastructure rehabilitation, and green infrastructure generally. When cities do this in common, pooling and thus increasing the impact of their purchasing power, the effect can be electrifying, lowering prices (via volume purchasing) and augmenting demand in the renewable energy sectors in ways that promote R&D as well as manufacturing. It can even add to urban jobs. Conditions can also be set that influence prospective vendors without coercing them.

3. *Carbon Tax:* The idea of imposing a tax on carbon to account for the environmental costs of fossil fuels (which economists call "externalities") typically not borne by either buyers or sellers, and that influences market competition, has a long history of support in Europe as well as bipartisan support in the United States. Congress, however, has recently refused to act on the idea despite its broad appeal. Cities and regions with taxing authority also have the option of employing the strategy. Given that cities are the primary users of energy, a municipal carbon tax could have considerable influence, especially in combination with other market and pricing approaches to fossil fuel usage.

4. *Fracking and Drilling Bans:* Cities are not primary venues for drilling or fracking, but few visitors to Los Angeles forget the

anomalous presence of small drilling rigs across the urban landscape. Around the world there are cities that sit on massive shale deposits where fracking is possible. When cities take action to prohibit fracking or drilling, as many have, the impact can be more than symbolic. Proof of the relevance of such bans is evident in the legal injunctions brought by state and national officials, often in seeming collusion with fossil fuel companies. New York State has banned fracking, and cities from Pittsburgh to Denver have passed urban prohibitions, but all have faced legal action. This is a domain in which common urban action is not just useful but necessary, and where a rights approach is strongly warranted.

5. *Waste to Energy:* Waste processing and disposal (elaborated below) is a universal urban problem that cities have traditionally responded to by dumping refuse in vast landfills that leak methane and CO_2 and absorb solar heat; or even arranging for remote disposal by exporting refuse to poorer nations willing to be paid to take on garbage. New technologies of waste incineration that can generate heat and biomass fuels and also capture methane offer a new, environmentally constructive response to the classic problem. The approach has been pursued in Scandinavia and is drawing attention in many cities, including Rotterdam and London (especially the borough of Southwark).

6. *Aboveground Mining:* Of all the novel ideas in play, perhaps none is as bold as aboveground "mining" for rare minerals, useful gases, and precious metals, a technique that can acquire valuable material resources for a city without requiring deep earth extraction and whether or not urban land sits on valuable deposits. This approach, described carefully by Australia National University physicist Penny Sackett and others, would reduce cities' dependence on mineral-rich regions. To take one example, phosphates can be

extracted from human waste (Mexico City's "black rivers") to manufacture commercial-grade fertilizer.

7. *Petroleum-Based-Product Regulation:* The universal use of plastics derived from the refinement of petroleum in water bottles, shopping bags, six-pack loops, and other environmentally destructive packaging is convenient for urban shoppers but an environmental scourge. Regulating such products through bans, bottle taxes, plastic bag fees and the like, along with the provision of renewable alternatives (such as biodegradable six-pack loops, glass bottles, reusable paper and canvas shopping bags) represents an easy and effective way to encourage environmentally sound behavior and reduce plastic manufacture and use. New York City has passed a plan to put a fee on plastic shopping bags, and small towns such as Lenox, Massachusetts, have banned plastic shopping bags.

8. *Built Infrastructure:* Energy wasted through heat and cooling loss in urban buildings is a high-cost environmental problem with an easy (though not inexpensive) fix: better insulation. There are many technically feasible options for cities to green their built infrastructure, from simple remedies such as painting roofs white to employing building materials with better insulation capacity (including wood structures and water walls, as well as construction materials predesigned for energy efficient disposal when the buildings come down) and the imposition of mandatory LEED (insulation) standards. Copenhagen's U.N. building, prominently visible from land and sea, is a model of green construction. Wooden skyscrapers, the use of recyclable and biodegradable materials, and high-rise residences that seem to sprout gardens are among the novelties that punctuate this intriguing arena of green design.

9. *Green Infrastructure:* Green infrastructure can encompass a broad spectrum of policies regarding efficiency and waste, energy

options, transportation, built infrastructure, and waste management. Here we use the phrase more narrowly, applying it (in M. A. Benedict's description) to a set of practices that utilize natural features to provide environmental and community benefits. It involves transforming unsustainable domains into green spaces. It can even entail a hybrid "green and gray approach" to (for example) managing storm water runoff.[2]

A host of small-bore practices can contribute to green infrastructure. To take examples from just one area, dealing with storm water and runoff, these practices include *downspout disconnection,* which reroutes rooftop drainage pipes to permeable areas; *rainwater harvesting,* using rain barrels, cisterns, or permeable land areas to collect rainwater; *rain gardens,* shallow, vegetated basins that collect and absorb runoff; *planter boxes,* urban rain gardens with vertical walls and open or closed bottoms that collect and absorb runoff; *bioswales,* sunken channels that treat storm water as they reroute it; *permeable pavements,* which treat and store rainwater where it falls; *green streets and alleys,* integrating green infrastructure elements into street and alley design; *green parking* and *green roofs,* covered with vegetation enabling rainfall water retention; *blue roofs,* designed without vegetation to retain storm water; and *urban tree canopy,* trees planted to reduce and slow stormwater infiltration by intercepting precipitation in their leaves, branches, and roots. These are just a few examples from one aspect of green infrastructure. A more elaborate discussion follows below.

10. *Transportation Infrastructure—Vehicles:* Given that cities are still organized around the internal combustion engine, the first aim of a green transportation infrastructure is to lessen the dependency on fossil fuel vehicles and private transportation. As Mexico City and other mega-cities in the Global South have made evident, this is a more than daunting challenge.[3] The development of light rail, above-

ground dedicated-lane bus lines, river and bay ferries, cable cars, self-driving cars, smart parking meters, electric-vehicle charging stations, congestion fees, alternative day access license plates, and high-occupancy-vehicle lanes are some of the more important options to break the dependency on the internal combustion engine and private automobiles as primary urban people movers. Yet there are no easy solutions: electric vehicles plugged in to an energy infrastructure tethered to fossil fuel only remove the problem to some distant smokestack. Buses and cable cars can discourage automobile use, but only if they are cheap, fast, and efficient—and built to withstand the heavy use they have drawn in cities such as Bogotá and Medellín. Jersey City recently completed a light rail system of the kind that, ironically, once defined urban transit. Los Angeles has extended a "metro" line to Santa Monica, a radical move in the freeways-forever City of Angels. And New York has proposed an interborough light rail system. But these are baby steps on a long, arduous journey to car-free cities.

11. *Transportation Infrastructure—Pedestrians:* The more radical approach to reforming transportation infrastructure is the creation of "walkable cities" where pedestrians and bicycles are favored over automotive vehicles. This is an easier option for new cities, or in canal cities such as Amsterdam, Venice, and Soujo where automobiles are impractical. Still, many older cities have tried to make urban space more pedestrian-friendly, and bicycles have played a salient role. A few developed cities such as Amsterdam have long relied on bicycles as urban transportation. But for most municipalities, the focus on bicycle transport is new, popular, and controversial. Bike lanes and bike-share programs can now be found in hundreds of cities on every continent, and in such unlikely places as clogged New York (where bike lanes encroach on already overburdened roadways) and Los Angeles (where great distances and crowded freeways make their use

problematic). Amsterdam actually has too many bicycles, more than the number of its inner-city population of 750,000. Fifteen thousand bikes are fished up from the city's canals each year, and tens of thousands more are locked to railings and abandoned, inadvertently forming a kind of permanent sculpture exhibition. Ironically, the greatest threat to pedestrians is not cars but reckless cyclists.

Sadly, rapidly developing countries on their way to economic dominion such as China are replacing bicycles with automobiles and trucks, missing the opportunity to leapfrog the automobile technology altogether by radically updating their bikes and creating a post-automobile high-tech bicycle economy.[4] On the other hand, European cities such as Copenhagen have introduced enormously successful and influential bicycle programs aimed at providing a safe, fast, convenient but also sustainable transportation infrastructure. Copenhagen's success is stunning: 88 percent of the population cycles to work. Symbolism can also play a key role: in his days as mayor of London, Boris Johnson liked to bike (if a little ostentatiously) to work across London Bridge. New York mayors have made riding the deteriorating subway lines there a gesture of green mindfulness as well as of working-class solidarity.

Bikes are only a piece of a successful walkable city. Green spaces are the goal: waterfront parks and piers, abandoned overhead rail lines converted to parkland (like New York's High Line project), streets and squares closed to traffic, created pedestrian shopping malls (as many European and especially German cities have done) and squares like Taksim in Istanbul, and even uncovered and reopened paved-over rivers to creating pedestrian-friendly public spaces (Providence, Rhode Island, and Seoul, for example).

12. *Energy Infrastructure—Local Generation, Time-Shifting, and Storage:* Producing energy locally, even if it is from fossil fuels, saves

costs of transmission and permits an eventual shift to renewable energy. It is also easier to store locally generated energy, permitting time-shifting—storing power generated in periods of low demand for use during peak times. Solar, wind, and tidal power are not available all the time, so storing this electricity to permit its use later is an important incentive for developing efficient battery and other storage resources. Aging transmission lines are a major problem nationwide and in countries around the world, where pushing electricity through old high-impedance wire can use up to half the energy being transmitted. Upgrading local transmission wires or building new ones addresses this problem directly if only locally.

13. *Energy Infrastructure—Renewable and Alternative Production:* Local energy production will allow cities to reduce their dependency on a fossil-fuel-powered national grid, or possibly even sever their ties to it, and instead invest directly in wind turbine, geothermal, solar, tidal and other forms of renewable energy production. Coastal cities can use offshore wind and tidal power, while solar and geothermal are available everywhere. Hydroelectric obviously works best in mountain regions, but power from rivers is a source for almost every city in the world. Heavy up-front costs for conversion can be challenging, but the investments yield long-term savings. The startup costs also give cities a motive to demand a greater share of their own tax revenues. Alternative and renewable energy is, to be sure, a contested arena even among environmentalists. Many refuse to consider the use of nuclear power despite recent improvements in safety. Others suggest that wind and solar are "intermittent sources" that can never fully replace fossil fuel generation, and that over the long term, only a reduction in energy use can seriously diminish greenhouse gas emissions.[5]

14. *Energy Infrastructure—EV Charging Stations:* Technological advances in electric vehicles and battery life have made EVs a realistic

alternative for private and public transportation, with a potentially enormous reduction in both fuel use and emissions. But the impressive advances that allowed Amsterdam to introduce two hundred electric taxis and Oslo to become a Tesla EV mecca have been inhibited elsewhere by a lack of EV charging infrastructure. Thanks to an increasingly dense charging infrastructure in California, EVs are rapidly spreading on the U.S. West Coast; yet they remain nearly invisible in the rest of the country. Tesla has its own charging stations, but they cannot be used by the cheaper and more popular electrics like the Nissan Leaf or the Ford Focus. The current limits on battery life have been the primary obstacle to expanded EV use. Ubiquitous charging stations could offset this significantly.

15. *Recycling:* Recycling is the most common and popular option for promoting a green urban environment. Collecting, recycling, and reusing such common materials as paper, plastic, metal, and glass can yield enormous environmental and economic benefits. This entails a public sector contribution, with regulation and collection stations both outdoors and in business and residential buildings. It also requires that the recycling process leads to the actual separation and recycling of appropriate items. Far too many urban programs are for show: collection stations offer deposit drops for paper, plastic, and glass, but these sometimes feed into a single common bin whose contents often end up in landfills! The great value of recycling is that it allows city residents to play an active role in green outcomes and to learn how individual behavior can bring collective goods. In many cities, the poor and homeless earn income by separating recyclables from garbage. Recycling has a significant environmental effect, and it also mitigates the impact of the overheated consumer economy on natural resources.

How Cities Can Work Together on Sustainability Policies

The approaches described above, most of which overlap, can be undertaken to great effect one city at time. They are also mutually reinforcing: networks of collaborating cities that pursue these strategies in common can amplify their global impact. Yet the principle of cities augmenting their influence through cooperation is much more than just a matter of numbers. Cities working together also make it more difficult for courts or regional and national governments to oppose urban environmental initiatives. Municipal collaboration can also affect markets through pooled purchasing or shared boycotts, giving private corporations an incentive to align their for-profit policies with urban public values.

Just as businesses, in the time before unions, had much more leverage to dictate the salaries they were willing to pay individual laborers, so today cities that act on environmental issues one by one are easily stymied by states that are resistant to local green policymaking. Around the world, states prohibit fracking bans, subsidize carbon, or compel cities to permit oil pipelines to run through town. But just as syndicated workers can bargain with employers in a common voice, cities united in urban networks can use a common voice to stand firm on common approaches to sustainability and decarbonization, together withstanding pressures from above they cannot resist alone. The impact of collective action obviously varies across cities. But at the most basic level, ten million citizens recycling in twenty cities beats two million recycling in one city. That is simple arithmetic. Geometric effects are possible in domains where one city is the same as none, but where concerted action produces significant results—in such contested domains as divestment or bans on drilling and fracking, and with pooled purchasing to lower market prices.

Many of these strategies are well known and widely appreciated, even if they are put into practice unevenly—recycling, bike-share programs, electric vehicles, or pedestrian malls, for example. Other approaches are familiar but controversial and hence subject to greater resistance: both the carbon tax and divestment attract movement environmentalists, but neither has been widely enacted. To set the table for cities aspiring to cooperate on sustainability, I want take a moment to elaborate on several innovative proposals that are less familiar but that have the potential to be effective in many different cultural, geographical, and economic contexts and can therefore provide a foundation for common action.

A number of these are driven by technology, whose potential remains underutilized. Technology responds to the market but also to the needs of the public sector (sometimes experienced as market demand—say for post-incandescent street lighting). Fortunately, cities are hubs of innovations and centers of creativity. Nearly all patents originate in cities, and university and medical research institutions are almost all urban. The contributions they make to turning the built environment or transportation environment or port environment into a green environment are crucial to sustainability. Many cities are cultivating technology incubators and sponsoring research institutions. The Massachusetts Institute of Technology's media and urban labs are working with cities worldwide. New York City has a private-public partnership in Brooklyn called ACRE (Accelerator for a Clean and Renewable Economy) designed to deal exclusively with clean energy production, while Rio de Janeiro hosts Morar Carioca Verde (Green Housing for Rio de Janeirans), which is piloting a program in the favelas of Babilônia and Chapéu-Mangueira as a laboratory for testing green practices that can later be applied to other communities. Barcelona has become a global urban laboratory, working with firms like Cisco Systems on transportation and communication and producing new networks such as Webnet that offer web-sharing for urban best practices everywhere.

The short list I will describe here includes surface "metro" style rapid transit bus systems, aboveground mining, waste-to-energy technology, and green infrastructure.

Surface Rapid Transit: Subway and underground rail systems are the most environmentally prudent and economically efficient forms of transportation a city can have, once they are built. But they require a heavy up-front investment and are dependent on topography and geology for their viability. Construction can also take years, even decades—witness the Second Avenue subway line in New York, first envisioned before World War I and not yet completed—during which surface transportation is endlessly disrupted by construction. These realities, daunting for Global South mayors without expansive resources, have led cities like Bogotá in Colombia to pioneer aboveground rapid transit bus networks that cost about five percent of what an underground metro costs yet carry large numbers of riders, including those commuting from distant suburban favelas, at speeds three times faster than ordinary bus service. In Medellín, a cable car and gondola system operates as long-distance urban transit, bringing poor workers from outlying hill regions into the city.

The secret to efficient bus transit is dedicated bus lanes closed to all other traffic (vehicles, bikes, pedestrians, cows, goats—not a joke!) by means of raised curbs. Bus stops are express and offer curb cuts for access only every three or four blocks. The benefits are economic as well as environmental. In much of the Global South (but also in, say, Paris) poor workers live in distant suburbs and must travel to the inner city to work. The round trip with normal bus service can take up to three hours each way, creating a work day that starts at four in the morning and ends at nine at night, with devastating effects on family life. When the trip is made more efficient, even middle-class residents often prefer to give up their precious cars in favor of rapid buses. The most serious problem for these buses has been such heavy use that they cannot be maintained properly to meet the demand.

Aboveground Mining: In an altogether novel and quite esoteric domain, cities have been given an opportunity to exploit "aboveground mining" in ways that can be enormously important for both sustainability and economic equality. In a pathbreaking article, the Sydney-based physicist Penny Sackett has written:

> The earth is finite and so are the chemical elements of which it is composed. Those elements, represented by the symbols that fill the boxes of the periodic table, fuel all human consumption. Yet we are mining and redistributing these fundamental elements at such a rapid rate that many are already in short supply or likely to become so in the next few decades. To maintain supply lines to the dinner table and to industry, we must completely reframe our understanding of mining, consumption, human environments, and waste, recognizing that the accessible elemental resources of our future are largely stored aboveground in the familiar objects of our daily lives.[6]

In spite of the vital relevance of the closed cycle use of physical materials for sustainability, a few countries dominate virgin source mining, requiring cities to buy and import these materials from distant vendors at a high cost. Imagine now that some of these precious assets can be mined aboveground locally in the cities where they are used. Phosphorus is an important example: key to food production and fertilizer, but currently available only from phosphate rock in a volatile market. Yet both phosphorus and nitrogen can be extracted from human waste, which every city has unlimited access to in the form of sewage. Other commodities can be mined aboveground as well, but they must first be surveyed and mapped. In Sackett's terms, this new approach offers the "possibility of changed consumer-producer relationships" and raises crucial questions about "who owns urban mines." (Aboveground mines are not to be confused with old-fashioned surface and strip mines, which get at minerals just at or below the surface and must be dug where actual mineral deposits are.)

When consumer items such as mobile phones, laptop computers, solar panels, and batteries are discarded as waste, it becomes possible to "mine" them for gold and silver as well as for indium—so widely used in solar panels that, ironically, the element is now on the "endangered" list. Replenishing the world's diminishing stocks of such precious elements through recycling can play a crucial role in urban environmental policy. Measuring how much lead, copper, and zinc (all "endangered" elements) is currently in use is difficult, but given their growing scarcity, subjecting these elements to aboveground mining and recycling makes economic and environmental sense for cities.[7]

Waste to Energy: In a story I recall with delight, Mayor Teddy Kollek of Jerusalem back in the 1980s was able to silence a noisy crowd of imams, rabbis, and Christian ministers arguing about access to Jerusalem's holy sites by declaring, "Spare me your sermons and I will fix your sewers!" Sewage is every mayor's problem—along with garbage, an even greater challenge. Indeed, nothing is more daunting or enduring on a mayor's agenda than waste disposal. Since waste is by definition a product of surfeit and inefficiency, it may seem counterintuitive to think of it as an element in the built infrastructure that can be made sustainable; or that there are reforms that are the equivalent of bike-share programs that can reverse its environmental effects. Waste streams need to be sorted and treated in terms of carbon density, recyclability for other uses (such as building materials), and costs (junk-mail disposal can cost ratepayers up to two million dollars a year!). Most challenging are waste materials with toxic disposal risks, such as electronics and batteries, or with adverse environmental post-use impacts such as plastic bags, now being proscribed from use in many cities, or carrying a charge to consumers.[8] Composting is a vital part of every waste management strategy. It reduces waste even as it increases the use of green space and biodiversity. It is the original netzero—a full-cycle supply chain.

But the most popular approach to waste management is the least environmentally efficient: relying on landfills, which (like fracking) produce the disastrous byproduct methane. Since the organic elements in waste decay under anaerobic conditions (no oxygen), landfill gases including methane are unavoidable.

This is where the new technologies associated with waste to energy (WtE) have become so salient. The scale of the problem is evident in the city of Ankara (the capital of Turkey), where the Mamak landfill serves a population of 3.6 million people and receives approximately 3,500 tons of waste material every day. Since 1980, more than 20 million tons of waste has been deposited on the Mamak landfill. To stem the escape of methane into the atmosphere, the landfill has been covered with foil to capture and collect the gas. That is a start. But the real aim must be to incinerate or decompose the refuse and use the resulting heat either to warm buildings or drive turbines generating electricity. In Ankara, the methane capture program mitigates up to the equivalent of 550,000 tons of carbon dioxide each year, which is equal to the CO_2 emitted by 160,000 cars each year. The method is costly, but it turns waste and its greenhouse gas byproducts into an economic asset.[9]

It is the great virtue of the new technologies that transform waste into energy that they not only promote sustainable development but offer an ingenious solution to waste disposal that does not depend on landfills or shipping refuse abroad. Waste-to-energy conversion by new methods of incineration is giving cities a powerful resource for manufacturing alternative energy. The burning of garbage yields two key products: heat, a direct form of energy that can be used to warm buildings or generate steam to drive electricity-producing turbines; and biomass fuels that, with (expensive) methane capture in place, can replace carbon fuel and reduce carbon emissions. In addition to incineration, there are new thermal technologies that rely on gasification, thermal depolymerization, pyrolysis of plastic back into oil, and anaerobic digestion and fermentation rather than direct combustion.

China, with more than fifty WtE plants, and Japan, with hundreds of plants, have been leaders in developing these technologies, although American companies such as Komar Industries and Bioenergy Corporation are making strides in fermenting the sugar in waste to produce ethanol. Esterification of feedstock has been the traditional means of producing biofuels, although many argue that in a hungry world, using grains to produce fuel is a poor tradeoff. This makes the use of waste far more attractive. Citywide programs are still rare, although Oslo is moving in the right direction; but one-off WtE plants are found in many cities across the world, for example in Brampton, Ontario; Stoke-on-Trent, England; Newark, New Jersey; and Fort Myers, Florida; as well as major cities like London, New Delhi, Vienna, and Vancouver.

There are environmental costs to WtE processes, especially when they are done on the cheap. The by-products of incineration can include carbon dioxide, methane, airborne and base carbon ash (what remains after burning), and such toxic chemicals as arsenic, all of which raise their own disposal problems. Capture, recycling, and deep burial of such residues can largely overcome the difficulty, but they add considerably to the up-front investment required. If WtE is to do more than shift the burden of toxicity and greenhouse gases to a later stage of the energy generation process, cities will need to make the investment—or explore the even more expense alternatives to incineration that are coming on line.

Green Infrastructure: We have seen that green infrastructure, in the generic definition employed by the U.S. Environmental Protection Agency (EPA), covers a broad spectrum of policies with respect to efficiency and waste, energy options, transportation, and even "built infrastructure" and "waste management" (which are sometimes viewed as rivals to green infrastructure and sometimes included in it).

The popularity of the green infrastructure approach is due in part to its breadth: its capacity to deliver multiple ecological, economic, and social benefits, and to perform multiple functions on a single land area. It responds to the diversity of urban environments in cities of every size and kind, and it incorporates conservation values and actions into land development, growth management, and built infrastructure planning.[10] Perhaps most important, it is concrete and specific in application, allowing mayors and city officials to grasp exactly what is required—and what is possible. Recommendations are robustly and usefully specific, as we saw in the discussion of rain and storm runoff above.

The many impressively specific measures for dealing with rainfall are examples of adaptation and resilience measures that address changing weather and climate patterns. But they do so in ways that are not always obvious; for example, by combating the urban heat island effect, in which temperatures rise in urban areas due to density, human activity, and dark constructed surfaces such as roads. Some researchers believe these heat islands influence global warming. One study of cities in China and India, although controversial, suggests that the urban heat island effect contributes up to 30 percent of global warming.[11] It undeniably contributes to the high costs of energy used by cities.

Plants absorb heat and carbon dioxide, provide shade, and increase solar reflectivity, known as the urban albedo, which refers to the potential of a given material to reflect sunlight as measured on a scale of 0 (black) to 1 (white).[12] New York and Los Angeles have introduced widely publicized programs to reduce the heat island effect, including painting urban roofs white, resulting in a radical increase in their reflectivity and a commensurate decrease in the heat retained by buildings.[13] Paris and other cities are following New York's lead. A U.S. Department of Energy study recently reported that "increasing the reflectivity of roof and pavement materials in cities with a population greater than 1 million would achieve a one-time offset of 57 gigatons . . . of CO_2 emissions. . . . That's double the worldwide CO_2 emissions in 2006."[14]

From Copenhagen and Seoul to Barcelona and New York, cities have scrutinized streets and plazas as well as rooftops, increasing tree plantings and converting sections of principal thoroughfares to pedestrian plazas (including the historical Ramblas in Barcelona and the recent installation of plazas at Broadway intersections in New York, including Herald Square and Times Square), moderating climate change but also enhancing the quality of life in the city.[15] Care must be taken, however, that quality-of-life arguments do not distract from the core environmental aims. When bike lanes or pedestrian plazas interfere with commerce (although this is more often a worry than a reality), environmentally sound changes can feel ambiguous, especially if quality of life is also an objective. Mayor Bill de Blasio's odd stand-off in New York City in 2015 over aggressive panhandlers badgering tourists for dollars, as the newspapers reacted with feigned horror at the spectacle of bare-breasted women mixing among the Disney characters and cartoon costumes in Times Square, put the new pedestrian mall itself at risk. Though many regarded this as a lurid diversion from real climate issues, the mayor's overreaction—he even suggested to his police commissioner that the city might dismantle the plazas and return them to vehicular traffic—was not helpful.[16] But it shows how quickly even the most committed public official can be distracted by the costly politics associated with a serious program of decarbonization.

Express bus surface transit programs, aboveground mining, WtE technology, and green infrastructure are just a few of the many initiatives open to cities seeking to address climate change. To be successful, they will have to seek such solutions together, which will allow them to pool resources and purchasing, act on key proposals in concert, and speak in one global voice to the national states and international organizations—bodies that should support them but too often stand in their way.

14
Exemplary Cities

Cities are powerfully positioned to act one by one to combat climate change and limit emissions locally. Given their overall contribution to global carbon emissions (they account for some 80 percent of the total), a collective effort by cities may be enough to effect full-scale global decarbonization. Individual cities' initiatives are limited only by the imagination and expertise of public officials and the political will of citizens. Since cities need not explicitly cooperate with other cities, they can acknowledge diversity and move to enact distinctive climate actions suited to their particular circumstances. From bike-share programs to aboveground rapid transit systems, green architecture and insulation to energy generation and distribution, we have already witnessed the extraordinary menu of choices from which cities can adapt policies suited to their distinctive size, stage of development, wealth, or geography. Even when working on stand-alone municipal programs, however, municipalities can model action for others.

Large metropolises such as London, Oslo, Seoul, and New York are just a few among many cities from Paris and Johannesburg to Bogotá and Boston whose development plans focus on sustainability as a core element of urban policy that integrates affordable housing, transportation, and green public spaces into programs that are inspirational for other places. Smaller cities like Bridgeport, Connecticut, have also made an impression.

London: Just sixty years ago, London's notorious use of coal for heating made it famous for pea-soup smogs. Even today when Beijing and Delhi are urban smog leaders, London is still one of Europe's most polluted cities, causing up to ten thousand premature deaths a year, according to the King's College Environmental Research Group. Yet Sadiq Kahn, the mayor chosen by Londoners in 2016, and the son of a Muslim Pakistani immigrant bus driver, responded to Great Britain's vote to exit the European Union—which not even his predecessor Boris Johnson anticipated—by doubling down on his green commitments. Demonstrating startling leadership, Kahn insisted that London had to have a voice in how Britain decoupled from Europe, especially on environmental matters, where Parliament might be tempted to water down tough E.U. standards. Speaking in the summer of 2016, on the sixtieth anniversary of the Clean Air Act (which those pea-soup fogs helped inspire), Kahn said he would not allow E.U. air quality rules, which have been poorly complied with in any case, to lapse. "Leaving the E.U. should not be the first step of us going back to being known as the dirty man of Europe," he said; instead the new government should be compelled to "put in place the strongest possible legal protections to ensure the existing legal limits are retained and not undone by Brexit."[1]

Kahn proposed a ten-pound "T-Charge" to be applied to the "most polluting vehicles" driving into central London, while extending the "ultra-Low Emission Zone" to the North and South Circular road and beyond by 2020. He also proposed scrapping diesel vehicles throughout England, introducing cleaner buses, and moving up dates on applying low-emission rules. Such measures, if successful, will not only limit the effects of Brexit on air pollution but contribute to decarbonization and reinforce London's exemplary role as the headquarters of the C40 Climate Cities. Kahn makes evident the difference a vocal and aggressive mayor can make in animating a city (and a country) to take on environmental challenges.[2]

Oslo: If London despite its role in helping to found the C40 Climate Cities under former mayor Ken Livingstone has been playing catch-up, Oslo has been in the forefront of sustainable urban development. With Norway's energy needs almost completely met by hydroelectric power, and its lion's share of North Sea oil and gas going almost entirely to exports—it is the E.U.'s most reliable source of natural gas—almost all of the income goes to sustain Norway's enormous Sovereign Wealth Fund. Oslo (with more than a quarter of the country's population of 5.1 million) has thus had the luxury of pursuing a zero emission campaign, and it appears likely to achieve that goal by 2025.

The zero emission target has been championed by Norway's Zero Emission Resource Organization, founded in 2002 to pursue a reduction of greenhouse gases through CO_2 disposal, wind power, biofuels, and renewables. It has also supported a chain of hydrogen fuel stations, known as the HyNor project, in and around the cities of Oslo and Stavanger. The Research Center on Zero-Emission Buildings has forged a partnership with the architecture firm Snøhetta to build a model carbon neutral "Multi-Comfort House" that (unusually for such designs) does not look like a windowless free-standing prison cell and can nevertheless be built reasonably cheaply. With well-insulated walls and built-in solar panels, the house will offset its energy uses over a period of sixty years, a formula that will make it even more efficient in southern zones with more sunlight. The house is an experiment intended to address the startling fact that 40 to 50 percent of carbon emissions in urbanized countries are the result of the construction and operation of buildings.

Oslo itself, with a plentiful supply of no-carbon energy, is applying the zero emissions goal with particular efficiency to the transportation sector. Visitors will be surprised to find a fleet of electric taxis dominated by what elsewhere are regarded as luxury cars, Teslas. Charging stations are plentiful, a crucial condition since charging infrastructure turns out to be the key to widespread use of electric vehicles. (This is one reason why California leads the United States in electric vehicle use, while in green-inclined New York City, usage remains negligible.) The goal is to make Oslo the most electric-vehicle-friendly city in the world, with a set of incentives that support converting all municipally owned cars to electric by 2015 (nearly achieved); all public transit to fossil-fuel free by 2020; all taxis to zero emissions by 2022; and close to 100 percent of new cars sold to emission free by 2025. One in four cars sold today in Norway are electric, while Oslo is also becoming a champion of green housing and architecture, with an especially strong example in its new opera house set in a neighborhood that gleams with green infrastructure everywhere around it.

Seoul: Asia also has exemplary green-leaning cities, including Hong Kong and Seoul. Seoul's mayor Park Won-Soon, currently the president of ICLEI, has put the environment front and center in his plans for development. As a small gesture pointing to the larger objectives, Mayor Park has cultivated a modest garden in his offices in city hall, where schoolchildren plant vegetables and can watch them grow.

The greater Seoul metro region has a population of almost 25 million, making it one of the world's five largest mega-cities. Thanks to the cumulative efforts of several administrations, it was ranked by the Sustainable Cities Index for 2015 as the most sustainable city in all of Asia. (Of the top ten, the first seven were European and the next three Asian. Not one U.S. city made the top ten list!) A considerable part of Seoul's success comes from a remarkable project begun by Mayor Lee Myung-bak in 2002. With unexpected public support (and some official resistance), he spent $900 million to remove an ugly elevated highway that had dominated the Jongno-gu district and restored the Cheonggyecheon stream, "re-daylighting the river that had been buried beneath it, and creating a spectacular downtown green space, all in under two and a half years."[3] The restored stream extends over six kilometers through the very heart of the city and is bordered now by two modest roads, a welcoming home to leisure events, recreation, and cultural happenings such as the Seoul Lantern Festival. As an observer has written in describing Seoul's green infrastructure, "environmentally speaking, the restoration of the *Cheonggyecheon Stream* has helped to increase wildlife in the area, cool down the urban heat island effect in the immediate vicinity . . ., decrease automobile traffic, and increase transit ridership. . . . It provides a corridor of open space in an otherwise extremely busy and crowded city center . . . both a welcoming getaway as much as a symbol for the future of a more eco-friendly Seoul."[4]

Other cities including Providence, Rhode Island, with its Waterplace Park positioned along the Woonasquatucket River, have also uncovered their rivers and moved highways in order to decrease pollution, increase biodiversity, ease traffic, and improve air quality—without diminishing commerce. Waterfronts, rivers, lakes, and seasides have played a special role in the renaissance many cities around the world have experienced in recent decades. By repurposing decaying industrial waterfronts as sites for habitation, recreation, culture, and green infrastructure, cities like Hamburg, Baltimore, London, and Barcelona, along with hundreds of others, have been made at once more attractive, more green, and more economically viable. The

Los Angeles River, much of it paved over in 1938, is the subject of a major revitalization plan including a Frank Gehry vision for its redesign that has inspired both hope and controversy.

Smaller, less obvious improvements in green infrastructure also count. Controlling invasive plants that undermine the health of native habitat can make a significant contribution.[5] Developing lanes and pathways conforming to nature's boundaries—so-called greenways—dedicated to non-motorized transportation, with bike lanes being the most obvious example, usefully addresses both climate and lifestyle. An influential American model is the East Coast Greenway, a project aspiring to create a network of trails for use by bicyclists and pedestrians from Maine to Florida.[6]

The Cheonggyecheon stream is not the only new green space in Seoul. On the banks of the River Han south of the city, Banpo Hangang Park was developed as part of the larger Hangang Renaissance Project. The park affords recreational space for Seoul's densely concentrated residents, with playgrounds, picnic locations, an inline skating track, soccer and basketball playing areas, bike lanes, and even a bike rental shop. Perhaps even more important than its contribution to climate, Banpo is also a pioneering exercise in flood control, where flood-prone flatlands alongside rivers are transformed into parks and recreational fields, with Cincinnati's Central Riverfront Park, and its Bicentennial Commons and the International Friendship Park, serving as the model example. It also acts as an environmental barrier curbing urban sprawl.

Near the airport to the south of Seoul is the new city of Songju, a green and smart "city of the future" that will quickly grow to two million people. Its plans include a parkland sewer system, universal broadband, integrated sensor networks, and green buildings to truly establish it as a model for the entire Seoul mega-city.

Like many Latin American and Asian cities, Seoul has made a massive investment in electric-powered public buses. It already possesses the world's third largest subway system, but its huge aboveground carbon fuel bus fleet of 120,000 vehicles has been a massive source of pollution and carbon emissions. Current plans are to convert half of this fleet to electric vehicles by 2020, which would be the world's most ambitious achievement of this kind if it can be reached. Seoul's smart transportation plan offers a pay card for local public transit that is integrated with private taxis and trains throughout Korea.

Finally, like many densely populated cities, Seoul faces the urban heat island effect, which can be remedied by a focus on green architecture and sustainable buildings (more than 40 percent of carbon emissions are the result of building construction and use). Urban density, human activity, and dark constructed surfaces and infrastructure can raise temperatures in ways that some researchers believe influence global warming. One study that measured the heat island effect in China and India suggests that it contributes up to 30 percent of warming, although this finding is controversial.[7]

New York City: No American city ranks in the world's top ten sustainable cities, and major American municipalities such as Boston, San Francisco, and Seattle are generally regarded as greener than New York. The Big Apple is more red, gray and brown than green, a rather old city with a rustic infrastructure, an ancient and deteriorating subway system, and a dense high-rise profile ill-suited to the amount of car and truck traffic it has to sustain. Its core, Manhattan Island, really is an island and, with its extraordinary density, might be a heat island as well if not for the vision of Frederick Law Olmsted, who in the early nineteenth century miraculously envisioned a "central park" in what at the time was still rural farmland.

Chaotic and exuberant, in the fashion of a city that never sleeps, New York nevertheless is accustomed to planning. Mayor Michael Bloomberg was only one of many mayors concerned with a habitable—today we say sustainable—city. During his three terms, he pioneered a broad effort under the name PlaNYC. More of it stayed on paper than he wanted, but an ambitious bike-share program and pedestrian plazas in the squares where Broadway intersects the major avenues became highly visible if controversial signature products of his urban imagination. Bloomberg's simple project of painting the city's acres of black tar roofs white to reflect solar heat became an exemplar to cities like Paris that embarked on similar programs. A white roof can radically increase a building's reflectivity and reduce the amount of heat it retains. A U.S. Department of Energy study has reported that "if all eligible urban flat roofs in the tropics and temperate regions were gradually converted to white (and sloped roofs to cool colors)," the effect on global temperature would be equivalent to the removal of "roughly 300 million cars (about the [number of] cars in the world) for 20 years."[8]

Under Bloomberg's successor, Mayor de Blasio, PlaNYC has morphed into OneNYC. Broader and (some think) a bit vaguer, it projects policy actions in domains from transportation and housing to community development and health care. OneNYC calls for the city to focus on three core goals: justice and equality, sustainability, and resilience. In all three domains it underscores community-based, bottom-up approaches. Its four engaging "visions" are described very broadly and not altogether modestly as:

Vision 1: "New York City will continue to be the world's most dynamic urban economy, where families, businesses and neighborhoods thrive."

Vision 2: "New York City will have an inclusive, equitable urban economy that offers well-paying jobs for all New Yorkers to live with dignity and security."

Vision 3: "New York City will be the most sustainable big city in the world and a global leader in the fight against climate change."

Vision 4: "Our neighborhoods, economy and public services will be ready to withstand and emerge stronger from the impacts of climate change and other 21st century threats."[9]

These are noble aspirations (and some might say unrealized boasts) and clearly will entail both major policies not yet enacted and the expenditure of major funding not yet appropriated. More important, the goals are unlikely to be achieved without active cooperation with other cities and networks. An earlier OneNYC report gives an enthusiastic account of the progress the city has made since 2011 in housing, parks, brownfields, and resilience, but does not begin to show how the visions will be realized. Indeed, it notes how much remains to be done with respect to air quality, solid waste, water, and energy. It sets additional and even more ambitious environmental goals, including reducing landfill waste to zero by 2030, expanding the composting program citywide by 2018, and adopting smart-grid technologies that will reduce transmission bottlenecks, enhance consumer consumption management, and spread peak energy usage across the clock. In 2016, the City Council tried to penalize the use of plastic bags (though it was thwarted initially by the state legislature) and has been active in lobbying for the state ban on fracking implemented by Governor Andrew Cuomo, with strong support from New York City–based environmental groups like 350NYC.

While it has a certain seductive resonance, the OneNYC plan is too new to be convincingly evaluated. It is a little short on deliverables, in part because its focus on climate justice and social equality risk watering down the strictly environmental goals. Yet an environmental plan that is not also an environmental justice plan is not only politically insupportable but morally untenable. The skewed distribution of the human costs of Hurricane Katrina in New Orleans as well as Hurricane Sandy in New York makes this all too obvious. In a disaster, the rich get out and then either rebuild or relocate. The poor go broke and become homeless. Or drown.

The most daunting and controversial issues in New York, as in many coastal cities around the world, relate to the comparative advantages of artificial preemption or natural resilience in the face of sea-level rise and storm surges. This question was dramatically illustrated by New York's vulnerability to Hurricane Sandy. With seventy-one deaths and $50 billion in damages, Sandy was one of the costliest disasters ever to strike the United States. Aside from the extensive damage it did in Staten Island and the Rockaways, it knocked out the subway system in Manhattan and left the borough without power below Thirty-fourth Street for days.

The New York debate is being reproduced in many places around the globe. Should cities depend on expensive and nature-averse approaches to storm surge, as New York did in proposing a multibillion dollar system of barriers and locks across the broad entrance to New York harbor? Or should they adopt more natural "resilience strategies" such as moving people and habitations away from vulnerable low-lying areas ("managed retreat") and restoring wetlands and marshlands that have traditionally protected vulnerable shorelines but have been removed in recent decades in the name of development?[10]

Global port cities such as New York, Seattle, Rotterdam, Hong Kong, Rio, and Los Angeles offer important lessons. The Port of Los Angeles, located in San Pedro Bay, twenty miles south of downtown, is the nation's busiest and is a crucial economic engine, contributing roughly $260 billion a year to the national economy. In March 2012, during the mayoralty of Antonio Villaraigosa, the port received its first ever gold-level LEED certification for its new Port Police Headquarters building. The Port Plan, however, went far beyond one green building and included a first-ever commitment to reduce goods movement to zero emissions and reduce greenhouse gases by 80 percent by 2050.[11] It now allows docking ships to turn off their idling diesel engines and power up from a customized port electrical system; and it requires the thousands of trucks that on- and off-load cargo daily to upgrade to hybrid engines with better fuel efficiency.

These and other measures have reduced port greenhouse emissions in Los Angeles by as much as half. Given that the port accounts for up to 40 percent of the city's aggregate emissions, these reforms are of real consequence.[12] Note that such efforts have succeeded in L.A. in part because they secured support from neighborhood forums and global port industry associations. The outcomes of the debates in New York and Los Angeles are likely to be critical to both sustainability and resilience in the world's cities, nearly all of them—global ports or not—fronting the world's waterways, whether rivers, lakes, seas, or oceans. The engagement and input of citizens will be crucial.

Bridgeport: It is not just big cities like Seoul and New York that wrestle with climate change. Bridgeport, Connecticut, with a population of less than 150,000, launched a program of environmental reform under the leadership of former mayor Bill Finch that made it a model in the U.S. Conference of Mayors and an Environmental Protection Agency regional showcase. Bridgeport has succeeded in removing six hundred tons of contaminated soils from an industrial site, invested a million dollars to provide low-income communities with access to waterfront parks, supported the certification of conduits handling wastewater and drinking water, and provided "greenscraper" training to local landscaping and contracting businesses to develop vegetated rain gardens to control storm runoff. Small-scale stuff maybe, but these measures prove that it doesn't take a megalopolis or a new smart transportation system to reduce emissions and raise the consciousness of a city's population.

Modest local efforts can catalyze funders and lead to larger projects. Bridgeport's efforts were rewarded with one of its largest federal grants ever, an $11 million TIGER grant (Transportation Investment Generating Economic Recovery) intended to help prepare the harborside for development by funding infrastructure improvements. The TIGER grant will be matched by an $18 million contribution by developer Bridgeport Landing LLC.[13]

Developing a menu of policy options for small cities like Bridgeport or large ones like New York or Seoul requires attention to both sustainability and resilience. Sustainability speaks to the capacity of policies to effect outcomes that are climate-friendly over the long term; resilience is more concerned to assure that where sustainable options are lacking, or fail to prevent dangerous outcomes, the consequences can be effectively and even creatively addressed. Judith Rodin and the Rockefeller Foundation's 100 Resilient Cities project, described in Part I, have forcefully demonstrated just how effectively cities can respond—or plan for action—when they stand at the intersection of sustainability and resilience; when, for example, they attach a "resilience officer" to city government and share practices and responses to the consequences of climate change with other cities. Resilience not only allows cities "to prepare for disruptions, to recover from shocks and stresses and to adapt and grow from a disruptive experience," but, when successful, offer a potent "resilience dividend . . . creat[ing] and tak[ing] advantage of new opportunities in good times and bad."[14] The approach of "managed retreat" is a robust example of resilience at its best.

At the same time, while resilience is an indispensable policy tool of cities, it needs to be coupled with sustainability strategies. Otherwise, we saw in Part I, there is a risk that it becomes an excuse for inaction in curbing climate change up front. For this reason I will continue to concentrate on sustainability, without forgetting that cities must understand readiness, responsiveness, and revitalization (the three crucial constructs in Rodin's prudent tool kit of resilience) if they are to survive in a world where climate catastrophe is a high probability.[15]

15

Trust Among Cities: An Index of Commensurability

The challenge is not merely to offer cities a menu of viable policy options that they can customize to their unique circumstances. We must also find ways to measure and compare their efforts through agreed upon standards. If cities are to work together across borders, as urban networks including the Global Parliament of Mayors propose, they must operate on level ground, where a wealthy city with a post-carbon economy—say Zurich, with hydro as an energy source, or Paris, with nuclear as a source—can compare its progress toward zero emissions with the progress of a developing urban economy still dependent on carbon—say Beijing, which largely draws its energy from coal. Otherwise we can expect neither justice nor political cooperation.[1] What constitutes a "sacrifice" for one city may be effortless for another. "We must run our air conditioners less" is not commensurable with "we must abandon coal as our energy source though we currently have no alternatives." And while air-conditioning is discretionary in Denver, it is lifesaving in Delhi.

For cities, the classic commensurability dilemma faced by nations is intensified: how to compare outcomes achieved, whether in efficiency, green or built infrastructure, or carbon reduction, when the parties are starting from such very different conditions of wealth, geography, stage of development, or demographics? The differences between poorer and wealthier, and Global South and Global North, are particularly salient. Developed cities managed to take their carbon energy benefit up front and at the expense of the rest of the world's cities (and nations), which now must pay the common costs of a warming climate for which the developing world bears far less responsibility. What today's "advanced" nations and cities wrought all happened long ago, before the science of climate change could begin to measure costs—before we even comprehended there would be any. Where does that leave generational justice? In the absence of compensatory transfers of wealth, the Global South will be neither willing nor able to give up overnight on a carbon dependency that the developed world took a couple of centuries even to consider abjuring. We apparently must relearn what the dialectic of "progress" means: that technical advances have costs, which are rarely distributed fairly.

The developing/developed dichotomy is only one contradiction. With so many policy options, cities must weigh all kinds of costs and benefits and calculate their offsets against each other. Constructing a dedicated express bus lane as Bogotá has, greening a port by electrifying ships at anchor as Los Angeles has, achieving a certain LEED rating, requiring hybrid public transportation vehicles as Oslo has, divesting from fossil fuels as Berlin has, and banning fracking inside city limits as Longmont, Colorado, has are all viable options: but how to evaluate their relative contributions to sustainability or resilience? Indeed, how much sustainability is the equivalent of how much resilience, if they can even be treated as commensurable? And as measured by what? The green infrastructure program of a Swiss city like Zurich already sourcing almost all of its electricity from hydro plants must be made commensurable with the green program of a French city like Lyons, which generates a preponderance of its energy from nuclear plants. And then compare them with a Polish city like Katowice or a

Chinese city like Chongqing sourcing their power primarily and (for the time being, necessarily) from coal.

Nations have faced this challenge for a long time. Their impressive lack of progress until just recently in securing a global warming protocol is as much about the perceived lack of fairness and commensurability among north and south, established and newly developing nations, as it is about inspection, monitoring, or sovereignty. Fairness is always a potential deal-breaker in contracts among parties with distinctive histories, characteristics, and interests. The terms of the latest U.S.-China pact are weighted down by these dilemmas, even if they are at the same time admirable because they manage to find common ground on fairness (or bypass fairness). Both parties agree that China's promise to *stop increasing* carbon emissions by 2030 is roughly equivalent to the U.S. pledge to *continue to lower* carbon emissions year by year, more or less at the current rate, until 2030. Fair deal? The parties say yes, and fairness is about perception. Yet what exactly is the common standard that permits one country to continue raising emissions for fifteen years and the other to continue to reduce them over the same period? And how might that standard be used to secure buy-in from other nations? These questions have flummoxed nations in all the meetings leading up to COP 21 in Paris. And the manner in which the questions were answered in Paris and reaffirmed in COP 22 in Marrakesh a year later does not instill confidence, since the standards depend on nations setting their own goals, reporting on their own progress, and taking their own sweet time.

There is only one way to achieve a standard of commensurability that makes fairness and thus cooperation feasible, whether among nations or cities: to establish mutually agreed upon indicators that allow comparisons across different sustainable policies and permit parties to an agreement to assess their distinctive contributions. How many downspout diversions are the equivalent of a bike-share program with three hundred bicycles? (Are the bikes free or rentals? Electric or manual? And do we measure aggregate numbers or per capita?) How to compare the green impact of a surface rapid transit bus system (Bogotá) with an overhead cable car system (Medellín)? And how then to compare their impact with respect to cars taken off the highways to their impact with respect to jobs?

These questions—it's complicated, folks!—bring us back to the search for broad common principles for measuring the value of specific policies. Cities are in effect being asked to develop a "green market"—public and private—without a common currency. Yet it is not just about devising an artificial currency, as the crisis with the Euro has proved. The challenge is to secure a currency that works for all individual members and is fair for all, one rooted in initial trust and capable of reinforcing and solidifying that trust through effective common practices that enhance both sustainability and resilience.

The good news is that there are a number of existing common indices around sustainable energy production and use, around decarbonization and green infrastructure, that point to the possibility of commensurability and thus fairness. The foundation for measuring sustainability exists.

The Leed Index: To get a feel for what a widely accepted comparative standard relevant to climate looks like, consider the LEED index (Leadership in Energy and Environmental Design). Although it is limited to one particular domain, energy usage and insulation in new construction and retrofitted buildings, it offers a model of commensurability, in this case, in evaluating different approaches to the built environment. Where green infrastructure is designed to mitigate climate change, the built environment is designed for many other purposes whose pursuit can often exacerbate climate change. In its current condition, built infrastructure can account for almost two-thirds of all greenhouse gas emissions.[2]

This not to say built environment and green infrastructure are mutually exclusive, even though they do reflect a perceived "green versus gray" narrative. That narrative itself invites innovative thinking about how to turn built infrastructure into a subsidiary of green infrastructure rather than its corrosive antinomy. Measures such as LEED that rate the energy efficiency of buildings help establish important metrics of sustainability by which various approaches to design and construction can be compared and evaluated. Many city building codes now incorporate the LEED standard, which applies to both commercial and residential construction and allows a U.N. building in Copenhagen to be compared with an office building in Los Angeles.

LEED has four levels, representing, on a scale from one to one hundred and ten, the minimum LEED certified (40–49 on the scale), LEED Silver (50–59), LEED gold (60–79), and the LEED Platinum (80–110).[3] A LEED Platinum rating can help a building sell, and thus makes economic as well as environmental sense. Having a scorecard allows for tradeoffs. Additional LEED credits in water management, for example, can offset credits lost in the choice of construction materials and insulation; or points lost in insulation deficits can offset points won through prudent water management. LEED has proven to be a logical baseline for so-called zero net energy (ZNE) strategies. Of thirty-two verified ZNE buildings in the United States by the end of 2014, fifteen, or nearly half, were also LEED Gold or Platinum certified.[4] Since then many more have been built or converted.

Copenhagen's new state-of-the-art United Nations building in its port neighborhood signals the city's commitment to achieving zero net environmental impact in all future construction. U.N. City, as it is called, is the regional headquarters for the United Nations and houses eight U.N. agencies and more than a thousand employees from one hundred countries.[5] It has a facade made from perforated aluminum shutters that can be regulated by employees to provide solar shading without interfering with harbor views, assuring a cool working environment not dependent on air conditioning; the roofing consists of recyclable organic materials that reflect sunlight and reduce solar warming; it also has 1,400 solar panels that generate 297,000 kilowatts of electricity annually, reducing demand from the city grid. The interior cooling system uses seawater from the Northern Harbor of Copenhagen, while the building itself employs a rainwater collection system that can capture 3 million liters of rainwater annually, cutting the building's demands on the city's water system.[6]

Green design obviously is more easily accommodated in new construction, but passive design in traditional structures can also be modified to accommodate environmental aims. The landmark seventy-seven-story privately owned Chrysler Building in New York City was retrofitted to achieve a Gold LEED rating through changes in its electrical systems and a series of small but important reforms affecting both electricity and water consumption, including real-time demand monitoring, outdoor air testing linked to calibration of the floor-by-floor air-handler units, and replacing original restroom fixtures with low-flow aerators, new water closets, and new urinals.[7] The road to sustainability is long; a great many short steps, however, can move cities down a road nations have difficulty traversing. Presidents talk grand policy; mayors install new urinals. Or get private companies to do it. Either way, the urinals turn out to make the difference in addressing climate change.

LEED is a real success but in a limited domain. What might a global LEED equivalent look like, not just for energy efficiency in buildings but for carbon use and emissions overall? Among existing indices that do what LEED does, but do it for greenhouse gas emissions more generally, two are worthy of careful scrutiny: the STAR Community Index and the Siemens Green Index. Without being diverted into an elaborate technical discussion, we can summarize how they work and their value for cities seeking commensurability.

The Star Communities Index: STAR (Sustainability Tools for Assessing and Rating Communities) offers a consensus-based rating system for community sustainability, providing a comprehensive set of goals, objectives, and performance measures by which different policies responding to different environmental, economic, and social conditions can be assessed. Founded by ICLEI-Local Governments for Sustainability, the U.S. Green Building Council, the National League of Cities, and the Center for American Progress, the STAR index helped create a community of leading local governments seeking long-term sustainability. The system is intended "to help communities identify, validate, and support implementation of best practices to improve sustainable community conditions."[8]

STAR addresses seven domains, including the Built Environment; Climate and Energy; Economy and Jobs; Education, Arts & Community; Equity and Empowerment; Health and Safety; and Natural Systems. It is not, however, a study in simplicity: its generic goals embrace forty-four objectives. The Climate and Energy generic goal, for example, includes such objectives as Climate Adaptation, Green House Gas Mitigation, Greening the Energy Supply, Industrial Sector Resource Efficiency, Resource Efficient Buildings, Resource Efficient Public Infrastructure, Waste Minimization. Much like LEED, STAR aims to transform the way local governments set and implement policy to improve performance. It favors a systems approach to sustainability where interconnectivity—call it interdependence—is paramount. Its credit point system assigns roughly equivalent points to each domain.

By using this point system, with certification levels ranging from three to five stars, cities can quantify and display their commitment to sustainability in ways other cities can recognize. Applying the STAR standard, Seattle earns a five STAR rating, Austin a four, and Atlanta a three. These ratings, however, are limited to the United States, for which STAR was developed, and they have not been tried internationally, where comparisons are always more problematic. Moreover, "the intent of STAR is not to rank cities across a standard set of indicators, but rather [to] provide a verified rating of their efforts given a menu of sustainability goals and objectives."[9] STAR's main use, then, is to offer a menu of options through which cities can improve.

The Siemens Green City Index: The Siemens Green City Index offers a complementary measure of comparison, made much more attractive because it is global. It is currently being used to measure the environmental performance of more than 120 cities worldwide. Siemens recognizes that different cities face different challenges, with their own politics, ecosystems, and citizen bodies. It offers a motivational tool by recognizing ecologically sound cities that outperform other cities by an objective measure, and that simultaneously offer lessons to underperforming cities seeking greener outcomes. The Siemens metrics and indicators are global in scope but achieve their universality by compromising their commensurability. The index is modified from region to region and continent to continent, so that its utility, if broader than that of STAR, is still circumscribed.

As one might imagine, being limited to North American cities with much in common, Siemens' U.S. and Canadian Green City index is especially nuanced, comprising sixteen quantitative and fifteen qualitative indicators in nine categories. It includes measures of

CO_2 emissions, energy usage, buildings, land use, transportation, water and sanitation, waste management, air quality, and environmental governance. Roughly half of the indicators within each measure are quantitative, usually based on data drawn from official public sources, for example, per capita figures for CO_2 emissions, water consumption, recycling rates, and air pollutant concentrations.

The kinds of comparison afforded by the Siemens Index are something less than fully objective, relying rather on qualitative assessment of a city's environmental policies. A municipality's commitment to sourcing more renewable energy, reducing traffic congestion, and addressing air quality are all taken into account as context, but these additions taint (or should we say strengthen?) quantitative measures with elements of the qualitative and subjective. Then again, political science is not hard science, and many important features of a city's climate policy cannot be quantified. The Siemens Index recognizes that defining what is important solely by what can be measured is a bad idea—like looking for lost keys not down a dark sidewalk where they were lost, but across the street under the lamppost where the light is better. Green intentions can hardly be quantified or compared to green outcomes, but given the role of political will in climate change, they surely are relevant.

Despite this "softness" and the absence of a simple quantitative ranking system of the kind the STAR index yields, the Siemens approach has achieved considerable legitimacy based on the organizations and institutions that have recognized it.[10] This credibility allows Siemens to enjoy a level of trust unusual for a private firm. But the index aspires to a universality it doesn't achieve. Its comparisons are most effective in assessing cities within specific continents and regions, but the greatest need for trust is among cities in different regions and continents that are radically different from one another in culture and development. Cleveland will be pleased to know how it compares with Seattle, but far more relevant to global climate policy will be how it compares with Cape Town or Medellín. North-South and East-West measures manifesting differences of wealth, development, and energy sourcing are those most likely to inspire distrust and thus most in need of agreed-upon standards of commensurability.

Within a country or region, the Siemens measure is both relevant and useful. The detailed indicators vary slightly from index to index so they can take into account the data readiness and the exclusive challenges in each region. The African index, for example, includes indicators measuring access to electricity and potable water, and the percentage of people living in informal settlements, reminding us that the very definition of a city—whether measured by absolute size or density of population—will differ on different continents. Although Siemens uses a common density scale to define cities, this scale starts in Europe at 10,100 persons per square mile, in Latin America at 11,700 per square mile, and in Asia at 21,100 per square mile.[11] If the meaning of the core term "city" is contestable, comparisons among "cities" will be problematic.

Not only are there slight differences in the measurement of North American and European cities, but Germany—Siemens's home country—has its own index distinct from the rest of Europe.[12] The three indices for Latin America, Africa, and Asia each have their subtle differences. With this much variety, it would appear that the Siemens Index is more helpful as a diagnostic tool (how are we doing on sustainability?) than a measure establishing full commensurability in evaluating sustainability policies in cities across the world. It allows cities to rank themselves as successful or wanting (and one hopes ready) to be more environmentally proactive, and is thus also (not surprisingly) a marketing and branding tool cities can use to boast about their success. Private-sector firms already use it to seek customers for green infrastructure materials.

Siemens uses categories and indicators that can be weighted and compared city to city, in accordance with a data index broken down into category indictors that are both quantitative and qualitative. Its weightings track its "normalization technique," which, for example, allows water consumption to be compared on the basis of cubic meters per head, from a minimum to a maximum based on common benchmarks.

The list is impressive, but data collection and measurement can be a greater challenge in some regions than others, which accounts for the utility of plural indices but also raises the problems of comparison and commensurability. Looking across the Green City Index regions, there are very few categories where a single data point—CO_2 emissions per capita, for example—is measured and reported in the same way in every region. Moreover, informal settlements, an ever larger part of the global municipal population with enormous impact on environment are often not included at all. No global index that does not account for the environmental impact of these burgeoning settlements is likely to be satisfactory.

What then can we conclude from this brief account of the LEED, STAR, and Siemens Green City approaches to rating sustainability? LEED leads the pack, mainly because it limits itself to one relatively standard measure, energy efficiency. STAR is intuitive and accessible, allowing public officials, NGOs, and ordinary citizens to understand more than they would from the metrics- and analytics-heavy Siemens approach. STAR is more detailed and more comprehensive and at the same time less complicated than Siemens. It is especially useful for letting individual cities evaluate their success in pursuing sustainability and show off their efforts. On the other hand, its virtues have a cost to commensurability, as it is not intended for comparing cities in measures of their progress.[13]

The Siemens Green City Index is better suited to comparison across cities, but not across regions and continents. Unlike STAR, with Siemens each city receives an overall index ranking as well as a particular ranking for each category. The results are presented numerically (for the European and the U.S. and Canada indexes) or in five performance bands, from "well above average" to "well below average." Although Siemens also employs qualitative criteria, its focus on hard data and details, which permits these numerical rankings and comparisons of what cities are doing or not doing, also gives it an academic feel.

Both the Siemens and STAR approaches suggest the limits of commensurability standards as sources of trust or as foundations for intercity cooperation. Fairness is the key to trust, but trust is also a matter of politics and perception, not just science and measurement. Civic judgments about intentions are as important as outcomes. Paradoxically, the trust that commensurability can help establish is needed to induce cooperation on commensurability in the first place. Without trust, strategies seeking to produce it are not viable. The bet here is that an index of commensurability, even if imperfect, will reinforce trust and make it easier for mayors to contemplate intercity cooperation and join urban networks or take the more ambitious step of joining a global urban parliament. But trust ultimately is associated with democracy, legitimacy, and the social contract. That is to say, with politics.

16

Realizing the Urban Climate Agenda

The good news about the effort to address climate change through government action is that it's happening. The bad news is that it's happening far too slowly. For every new hydroelectric plant built in the Global North, some enormous lake dries up in the Global South—the glacial lake Riesco in Chile and Bolivia's second largest lake, Poopó, both literally vanished over the past few years. California is a national leader in green public policies and regulates water usage scrupulously, but farmers there also grow pecans in semi-desert conditions where each nut harvested uses up to three hundred gallons of precious public water. For every urban fracking ban enacted, there is a move to block it in the courts. Coal is shut down, but fracked natural gas is accepted as a "transition" fuel—however pernicious its effect on decarbonization. And just as American environmentalists begin to have an impact on the fracking of shale, Europe proposes drilling its much less accessible shale reserves to limit its dependency on Russian natural gas. The best-case scenario for what is likely to be done through COP 21, the United Nations, and parallel worldwide nation-based environmental programs hardly dents the worst-case scenario for the catastrophic consequences of all that is not being done. New limits on HFCs and new green commitments by carbon-producing behemoths like China and India help, but the victory of climate deniers in the U.S. presidential elections makes them less than decisive.

These discrepancies are why cities matter, why urban politics counts, why *glocal* action that turns local governments into agents of global change is our best bet for making up the gap between our good will on climate change and the limits of state action. We can glean from the discussion here that there is an ample menu of sustainable options available to cities wishing to address climate change aggressively and that they can amplify their impact by coordinating their policies. Even in the absence of a definitive commensurability index, there are established indices that offer some degree of commensurability to cities wishing to cooperate. One way or another, cities must eventually develop such an index, perhaps one that borrows elements from LEED, STAR, and Siemens.

The infinite variability of cities will always make commensurability challenging and can leave cities with doubts about fairness and hence about working together, even as they endeavor to do so. But there are a number of ways to bolster trust by focusing on what cooperating cities have in common, even where measures of commensurability are found wanting. Among these commonalities is the experience of urban networks in seeding collaboration. While a commensurability index would facilitate cooperation in bodies like the C40, ICLEI, or the Global Parliament of Mayors, those bodies themselves, convened in the absence of such an index, may develop the trust upon which commensurability standards ultimately will depend. A renewal of the social contract that envisions cities as primary institutions responsible for securing key sovereign goods—life, liberty, and sustainability—can itself inspire the trust of urban citizens in concerted action and common policy making. Democracy is founded on trust: a social contract is a promise by the participating parties to be faithful to the obligations and responsibilities it establishes in the name of the rights it guarantees. Trust is the currency of the social contract.

Even if perfected, then, the social science of commensurability will no more automatically yield reciprocal trust among cities than the science of climate change has yielded a common politics of decarbonization. As I said at the opening of this book, the challenge is the politics, not the facts. The politics of trust tracks the politics of democracy. Democracy today has more resonance locally than nationally because there is an emerging consensus that cities everywhere represent the common will more effectively than any other governing body.

The last lesson of this study is then identical to the first: outcomes depend on politics as much as science. The challenge facing cities and citizens is to summon the needed political will to do the things we know how to do but have not done, and then to do them democratically. That will not be easy because democracy is in trouble, because moneyed interests and global oligarchies are corrupting government, because citizens are growing more cynical, alienated, and angry, and because cities still have not figured out how to pool power and pursue common goods as sovereignty trickles down to them. But even a skeptic about city power such as Richard Schragger agrees "what is required" to challenge the forces blocking the empowerment of cities is "a political movement."[1]

We end where we started, with the fate of the campaign against climate change and other existential threats dependent on democratic politics within and among cities.

No one needs to read this book to be convinced that cities are the coolest political institutions on earth. Chances are roughly two to one or better you live in a town or city, and not just for economic reasons. Spend a few days in Singapore or Cape Town or Seoul or Nashville. Witness Olso's cool Tesla taxicabs or Seoul's rehabilitated center-city river or Medellín's public cable car system or the green dreams of

Paris portrayed on the cover of the book in your hands. In this new millennium, keen to confront global warming but not yet fully empowered and resourced to so, cities are contenders and their citizens players. If nothing else, however, this book should also make clear that empowerment for cities must entail collective self-empowerment and that urban responsibility for global problems arises out of a sovereign default that prevents ever more dysfunctional nation-states from acting. This means that cities not only must accept the responsibility for assuring a sustainable world but must assert their right to do so. That is why the Global Parliament of Mayors conceives of itself in its Mission Statement as a "global city rights movement."

There are two formidable obstacles blocking a larger role for cities: a paucity of resources to take on new responsibilities—call it the *unfunded mandate question.* And closely related to this question, the absence of autonomy and jurisdiction in the face of the continuing (if faltering) claim of nation-states to a sovereignty they are no longer able to exercise fully but are not ready to relinquish—call it the *sovereignty question.* The unfunded mandate and sovereignty questions undermine the capacity of cities not only to address global problems but to deal with their own challenges, which are often a local reflection of universal problems. This is true whether cities are being constrained to repair their own infrastructures (sewers and water lines, for example), to host and house floods of refugees admitted by national authorities but quickly turned into a municipal responsibility, to combat terrorism which targets cities almost exclusively, or (our focus here) to take meaningful action on sustainability and resilience when central governments fail to act.

To put it a little more polemically, the truth is that far from ruling the world, mayors today sometimes feel they are scarcely able to govern their own municipalities—and that when they try they often have neither the resources nor the authority to do so. Empowerment is the remedy; better yet, self-empowerment. What they are not given in the way of resources they may have to impound from the revenues their productivity generates (80 percent or more of global GDP). Where their jurisdiction is denied by "higher authorities" they may have to remind those who think themselves "higher" that cities and their citizens have their own fundamental legitimacy, resting on the will of the people and the implicit social contract that the people's urban citizenship manifests. If their municipal autonomy is insufficient to the tasks cities must undertake in cooperating with others globally, they may have to declare a certain independence from the nations to which they are "subsidiary." This is less a declaration of urban *independence* from the nation-state than a declaration of urban *interdependence* among the world's cities, embracing their need and right to engage in common democratic work.

Cities have the right to govern where states can't or won't, but they do not yet have the power to fully exercise this right. Since right precedes and legitimates power, that is perhaps as it should be. Still, the absence of power does suggest, as Gerald Frug has written all too aptly (if a little too cynically), that "the more useful title [for my earlier book *If Mayors Ruled the World*] would be 'If Mayors Ruled Their Cities.'" All too true, Professor Frug. States may no longer know how to exercise their sovereignty, but they know how to prevent cities from doing so. Mayors must prove their ability to govern their municipalities locally before they can defend the social contract globally. National courts, steeped in national law and taking for granted the validity of the traditional nation-state social contract, are obviously unlikely in the first instance to side with cities. When Washington, D.C., tried to ban handguns, the Supreme Court denied its right to do so in the landmark *District of Columbia v. Heller* decision, ruling 5–4 that the Second Amendment protects an individual's right to possess a handgun. When Longmont, Colorado, banned fracking, the state courts stayed its action (although they also stayed further drilling by the fracking company while the case was appealed). In unitary states such as England and France, cities do not enjoy a discretion to act commensurately with their responsibilities and needs. Even in federal systems such as Germany, many key functions, including the police power, are vested in the *Länder* (equivalent to provinces or American states).

Even when higher jurisdictions favor (or at least refrain from obstructing) municipal action across borders that addresses global challenges, they often do not fund the mandates they lay on willing but overburdened and underfunded municipalities. National politicians are forever urging cities: "Go ahead: you can do it! Fix your infrastructure! Convert to alternative energy! Build resilience!" And then add, "Just don't ask us to pay for it." I keep repeating that cities produce four-fifths of the wealth from which central governments draw their revenues, but those same proprietary governments do not give the money back when cities are forced to act in the nation's stead.

Among cities' myriad grievances, nothing is more heartfelt or widely shared than the complaint about raising the lion's share of revenues collected by state and national governments and getting back only a lamb's share of the proceeds. Responsibilities are heaped on cities by dysfunctional higher jurisdictions, but the grants covering those responsibilities are rarely adequate to what is being asked. The issue of fairness we have focused on—fairness among cooperating cities—is eclipsed by the issue of fairness between cities and the states to which they are seen as subsidiary but whose responsibilities they increasingly must assume.

The Global Parliament of Mayors, which held its inaugural session in September 2016 in The Hague, is a momentous if partial step on the road to urban empowerment. For cities in Europe, deploying common power and enacting common policy expresses an implicit right recognized by the Council of Europe's Charter of Local Self-Government from 1985. That charter, to which little attention has been paid, "commits the ratifying member states to guaranteeing the political, administrative and financial independence of local authorities."[2] It goes on to say that "direct cooperation with individual local authorities of other countries should also be permitted." In its 2016 regional meetings, the European Union's Council on the Regions and Cities underscored the reciprocity of relations between the E.U. and cities.

This history fortifies the key premise of the Global Parliament of Mayors, which according to its Mission Statement is a "global city rights movement." Cities have the right to collaborate across borders. As Michael Bloomberg wrote in 2015, "central governments are not quick to devolve power, but they are doing so with greater frequency as they recognize the national benefits that come with local control."[3] The European Charter itself puts decentralization at the center of its purview, stating in Article 2 that "the principle of local self-government should be enshrined in written law." Witnessing recent trends in the United Kingdom and the United States and also in the U.N. secretary general's inclination to favor a decentralizing approach to cities, Bloomberg states the obvious when he writes that the "trend will only accelerate as the world becomes increasingly urbanized and cities become increasingly connected to one another."

The real question is whether cities will gain the cooperation of national governments in taking on global burdens; and whether those governments can acknowledge (if only implicitly) the growing disconnect between the sovereignty defining their independence and the reality of global interdependence, and hence accept a gradual devolution and decentralization to local bodies capable of global collaboration in the interests of all. Toward the end of his term, Secretary General Ban Ki-moon excoriated the U.N.'s more powerful nations, above all those on the Security Council, for their sclerotic inaction in the face of global crises. "You're not doing much at all," he complained.[4] Cities cannot fill the peacekeeping void, but there is much they can achieve when funded and empowered on other fronts. After all, the good of citizens is the same whether they identify with cities, provinces, or national governments—or conceive of themselves as cosmopolitan citizens of the planet. They aspire to secure property and some modicum of liberty. They seek sustainable lives for themselves and their children. It is of less consequence to them at what level of administration this happens, than that it happens. If sclerotic nations and the international bodies composed of nations are no longer up to the task, citizens will turn to their cities, with their ancient pedigrees and their modern know-how.

There are three political steps that must be taken to assure that a Global Parliament of Mayors and the other urban networks can succeed in securing justice and sustainability for their citizens. Each step depends to a degree on the success of the previous one. The first step is for cities to acquire (or retain) the resources necessary to govern in the expanding policy domains for which their right to guarantee their citizens sustainable lives makes them responsible. Second, cities must establish their right to self-government collectively beyond national borders, hopefully with but if necessary without the permission of national governments. Third, cities have to organize and sustain an urban rights movement and perhaps, to be relevant in the national politics of their home countries, generate a global rights-based

"Urban Party" that permits their citizens to lobby their regional and national governments to accede to their just demands for autonomy, resources, and legitimacy.[5]

Most political movements today derive their persuasive authority from nineteenth-century ideologies of state and market, or community and individual, in ways that deprive cities of their natural majoritarian constituency on behalf of urban values and cosmopolitan goods. Cities throughout the world tend to be more "progressive" and cosmopolitan than nation-states, which remain ideologically rigid and to some degree more beholden to rural and suburban than purely urban interests. And even cities have been subject to takeover by left populist parties less interested in cosmopolitan values than in anti-elite ideology: Podemos in Spain, which controls city halls in Barcelona and Madrid, or the Five Star Movement in Italy, which not only won control of Rome and Turin but helped defeat Prime Minister Matteo Renzi's constitutional reforms at the end of 2016. His plan to reform the Senate to represent metro-regions went down to defeat, and Renzi himself resigned.

Yet even where cities do pursue urban goods and urbane values, they rarely influence the outcome of national elections to the degree their numbers suggest they should. The European Union still seems to favor regions over cities, and to work more on agricultural subsidies than affordable housing in its cities. In the United States, the structure of congressional (particularly Senate) representation means that a suburban and rural minority dominates the legislature and rules over the urban majority, trends that are reinforced by gerrymandering and voter suppression activities.[6] In other countries, such as Thailand, differences between urban and rural parties are undermining the commitment to democracy itself.

This is by no means the only reason why majorities are deprived of their full voice. The failure of urban leadership to identify a true urban agenda is a contributing factor. It would require another book to develop the full argument for a politics of cosmopolitan cities organized around the common pursuit of global goods. Suffice it to say that cities today invert the traditional notion of local politics as parochial and national politics as cosmopolitan. In the United States, it is cosmopolitan cities that have championed gay marriage, LGBT rights, identity papers for undocumented immigrants, gun control, minimum wage laws and both affordable housing and pre-school education policies. And it is a gridlocked and parochial national government that once purported to represent "universal values" that has often stood by mutely or even actively opposed urban cosmopolitanism and its global goals. In Great Britain, the English public voted for "Little England" and an exit from the European Union, while urban London voted overwhelmingly in favor of remaining. Some cities in the north such as Bury, Blackpool, and Birmingham, however, voted to leave, perhaps identifying more with archetypal rural England than with its great capital in the south.[7]

Because urban citizens are the planet's majority, their natural rights are endowed with democratic urgency. They carry the noble name of "citizen" associated with the very word *city*. But their aim is not to set urban against rural, or replace the current dominion of suburban and rural constituencies over urban constituencies with the dominion of urban over rural. It is to restore a more judicious balance between them by giving the urban majority its due and assuring its rights and interests are not constantly violated by a non-urban minority. It is also to foster metro-regional associations such as Grande Metropole Paris or Greater Detroit that allow city, suburb, and countryside to see what they share (the countryside's food, water, and green space but also the city's culture, transportation networks, and jobs) and how much they need one another. This rebalancing is more important than ever in a volatile period of reactionary and divisive populism during which, despite two and a half million more votes cast for Hillary Clinton in the United States, Donald Trump won a convincing electoral college victory.[8]

Once upon a time, nations aspired to universality, and local jurisdictions were parochial. With populism on the rise and nation-states mired in dysfunction, the valence is reversed. Today it is cities that look forward, speaking to global common goods, while fearful nations look back, ever more parochial and xenophobic. Urbanity is a global virtue associated with tolerance, creativity, interdependence, and multiculturalism. It seems attuned to history, if not destiny. Nationalism, seeking to restore a sovereignty that has not been stolen but irrecoverably lost, has a narrow compass defended by walls that don't work. In an Urban Party coalition, cities can take up the cause of glocal goods, goods that are necessarily sustainable. For sustainability, which is synonymous with survivability, is the alpha and omega of all relevant policy in the Anthropocene, not just for city dwellers but for all.

The world is getting too hot. Science makes it clear that sustainability is both necessary and possible. Politics shows it is achievable. Cities are poised to make it happen, which is what makes them so cool.

Notes

Introduction

1. Donald J. Trump, Twitter message, @realDonaldTrump, January 1, 2014 (https://twitter.com/realdonaldtrump/status/418542137899491328). As with his race and gender comments, Trump said baldly and impolitely about climate change what many Republicans have said politely in earlier times, as the comments cited in the text below by Ted Cruz and George H. W. Bush show.

2. Oliver Geden, "The Dubious Carbon Budget," *The New York Times,* December 1, 2015.

3. Coral Davenport, "Nations Agree to Cut Use of a Harmful Coolant," *The New York Times,* October 6, 2016.

4. Figures from Elizabeth Kolbert, "The Siege of Miami," *The New Yorker,* December 13, 2015.

5. Bill McKibben, *The End of Nature* (New York: Anchor, 1989), p. 39.

6. William D. Cohan, "To Save the Planet, We Need to Leave Fossil Fuels in the Ground—but Oil Companies Have Other Plans," *The Nation,* June 29, 2015.

7. Samuel Thernstrom, "The Next Shale Revolution?: The Astonishing Promise of Enhanced Oil Recovery," *The Weekly Standard,* December 29, 2014.

8. Naomi Oreskes and Erik Conway offer a powerful account of how doubt can be evoked, whether about smoking or CO_2 emissions, in *Merchants of Doubt: How a Handful of Scientists Obscured the Truth on Issues from Tobacco Smoke to Global Warming* (Bloomsbury, 2011). The book was made into a compelling film by Robert Kenner in 2014. The book's argument that such campaigns are less good-willed expressions of scientific solidarity but obfuscations rooted not in outright denial but in "scientific"-seeming doubt was reinforced by the news that Wei-Hock Soon, a scientist who challenges global-warming science, has been paid more than $1 million by organizations in the fossil-fuel industry. Justin Gillis and John Schwartz, "Deeper Ties to Corporate Cash for Doubtful Climate Researcher," *The New York Times,* December 21, 2015.

9. McKibben, *The End of Nature,* p. 51.

10. I have addressed this dilemma directly in my essay for the Center for Humans and Nature called "Democracy or Sustainability?: The City as Mediator," *Minding Nature* 7, no. 1 (January 2014).

11. At this writing the term *Anthropocene* has not yet been ratified by the International Geological Society but that body seems poised to give it official status.

12. Julian Castro, quoted in Liz Enbysk, "Julian Castro: Cities Are Where the Future Happens First," *Smart Cities Council,* September 2015 (http://smartcitiescouncil.com/article/julian-castro-cities-are-where-future-happens-first).

13. Michael Bloomberg, now special assistant to the U.N. Secretary General for Climate Change, "City Century: Why Municipalities Are the Key to Fighting Climate Change," *Foreign Affairs,* September/October 2015, pp. 116–124.

14. Pope Francis, *Laudato Si': Care for Our Common Home,* Encyclical on Climate Change and Inequality (Brooklyn: Melville House, 2015).

Chapter 1. The Social Contract and the Rights of Cities

1. The Hague Declaration, promulgated at the founding convening of the Global Parliament of Mayors, September 10, 2016, Appendix 20. The declaration also affirms that it "builds upon the aspirations and achievements of the United Nations, the OECD, COP 21, Habitat III, The 'Inclusive growth Mayors Network,' The World Bank and the many vital urban networks already addressing these challenges, including the C40 Climate Cities, ICLEI, Eurocities, The U.S. Conference of Mayors and many others."

2. Richard Schragger, *City Power: Urban Governance in a Global Age* (Oxford University Press, 2016), p. 5.

3. The Ninth Amendment reads: "The enumeration in the Constitution of certain rights shall not be construed to deny or disparage others retained by the people." This formulation manifests the federalist principle of vertical separation of powers, and the ultimate source in the people of rights claims. The Tenth Amendment reads: "The powers not delegated to the United States by the Constitution, nor prohibited by it to the States, are reserved to the Sates respectively, or to the people."

4. The United States Conference of Mayors, Council on Metro Economies and the New American City, June 2014.

5. Ibid. These comparisons are with American cities, but comparable statistics are found around the globe.

6. These are rough numbers intended for exemplary purposes; estimates vary depending on whether the source is the World Bank or the Brookings Institution or the United Nations.

7. John Schwartz, "Climate Change Reaches Courts as Citizens Sue," *The New York Times,* May 12, 2016.

8. Working with cities in organizations like Habitat III, the U.N. has pivoted to cities as action agents.

9. Schragger, *City Power,* pp. 258–259.

10. Thanassis Cambanis, "Beirut Upstarts Find Footing in Political Quagmire," *The New York Times,* May 11, 2016. Ziad Baroud, a former interior minister, said of the Madinati movement: "They are well organized, and they have content . . . they are into politics, that's why they made it."

Chapter 2. The Devolution Revolution and the Politics of COP 21

1. The COP 21 agreement was signed later, on Earth Day 2016, when the required fifty-five nations, representing 55 percent of emissions, put their signatures to the accord it and made it official.

2. "Historic Paris Agreement on Climate Change," U.N. Climate Change Newsroom, December 12, 2015, Paris.

3. Justin Gillis, "A Path Beyond Paris," *The New York Times,* December 1, 2015.

4. William J. Broad, "Satellite System Would Track Individual Countries' Emissions," *The New York Times,* May 10, 2016.

5. The "Global Protocol for Community-Scale Greenhouse Gas Emission Inventories" is described in details on the Web site of ICLEI, Local Governments for Sustainability (www.iclei.org).

6. Coral Davenport, "Obama and Chinese Leader Vow to Sign Paris Climate Agreement Promptly," *The New York Times,* April 1, 2016.

7. See Benjamin R. Barber, *Strong Democracy,* 20th anniversary edition (Princeton University Press, 2004); and the new collection of essays *Strong Democracy: Promise or Peril?* ed. Trevor Norris (Lexington, 2016).

Chapter 3. Climate Change in the Anthropocene

1. Following philosophers like Rousseau and Marx who could portray modernity's benefits and costs in relationships of reciprocal contradiction and resolution, Max Horkheimer and Theodor Adorno offered what is perhaps the classic portrayal of the two-sided face of progress, in *Dialectic of Enlightenment,* ed. Gunzelin Schmid Noerr

(Stanford, Calif.: Stanford University Press, 2002; originally published in 1947). The Anthropocene is the Enlightenment dynamic written into the handbook of geology.

2. John Rockstrom, Will Steffen, et al., "Planetary Boundaries: Guiding Human Development on a Changing Planet," *Science*, February 13, 2015 (http://science.sciencemag.org/content/347/6223/1259855). Jeffrey D. Sachs has put the idea of planetary boundaries to good use in *The Age of Sustainable Development* (New York: Columbia University Press, 2015), pp. 81–92.

3. Bryan Walsh, "Ocean Acidification Will Make Climate Change Worse," Time.com, August 2013 (http://science.time.com/2013/08/26/ocean-acidification-will-make-climate-change-worse).

4. Sachs, *The Age of Sustainable Development.*

5. Aboveground mining is one of the more dazzling technological innovations, introduced by Penny Sackett. See Sackett, "Elemental Cycles in the Anthropocene: Mining Aboveground," Special Papers (#520), Geological Society of America. A shorter version appeared in *Solutions Magazine* 3, no. 3 (June 2012).

6. Edwin Black, *IBM and the Holocaust* (London: Little, Brown, 2001).

7. See Judith Rodin, *The Resilience Dividend: Being Strong in a World Where Things Go Wrong* (New York: Public Affairs, 2014). I explore the resilience approach more fully in Part II.

8. It has always concerned me that alongside the careful city-preserving strategies of programs like Rockefeller's 100 Resilient Cities can be found an easy and imprecise discussion of resilience by carbon energy companies for whom it is a convenient way of accepting the consequences of continued carbon consumption rather than finding ways of avoiding them. A good way to assure that resilience is part of a sustainability strategy is to link the two terms *resilience* and *sustainability* and focus on urban policies that occupy the intersection between them. It is hard not to be skeptical when a company such as Royal Dutch Shell organizes and promotes a major conference focused on resilience alone, as it did in Rotterdam in 2014 (at which I spoke).

Chapter 4. The Facts Are Mute, Money Talks

1. Jeffrey Sachs, remarks at the Harvard Club, March 31, 2015.

2. Thernstrom, "The Next Shale Revolution."

3. Richard Berman, cited by William D. Cohan in his aptly titled "Big Oil Wants to Burn It All," *The Nation,* December 29, 2014.

4. Although it received permission from the Obama administration to drill in the Arctic Ocean, Shell abandoned its exploratory mission for technical and market

reasons: it was too difficult to drill in the region and the price of oil was too low to justify the cost. It is still about profits, not science. It can be done, but it's just too expensive to be worth it.

5. Harold Lasswell, *Politics: Who Gets What, When, and How?* (New York: Whittlesey House, 1936).

Chapter 5. Privatization and Market Fundamentalism

1. For a brilliant account of this technique, see Oreskes and Conway, *Merchants of Doubt.* On the tobacco lobby, Allan Brandt's book *The Cigarette Century* is revealing.

2. Thomas L. Friedman, "Trump and the Lord's Work," op-ed, *The New York Times,* May 4, 2016.

3. There is a simpler approach to campaign finance reform than trying to regulate the amounts of money raised and spent or overturning the *Citizens United* decision. It is to restore the public airwaves to their purpose as a public utility as described in the Federal Communications Act of 1934. There is no good reason that the airwaves cannot be restored to the public use during elections to offer a free and level playing ground for all candidates (who spend up to 70 percent of funds on broadcast advertising). Why should citizens have to buy back publicly owned broadcast airwaves by paying private media companies (who have leased the public airwaves) to use them during elections?

4. See Frederic C. Rich, *Getting to Green: Saving Nature—A Bipartisan Solution* (New York: W. W. Norton, 2016).

5. Such market fundamentalism has a long tradition in political science, going back much further than Margaret Thatcher. In the early 1950s, a seminal book by David Truman denied the very existence of a public interest, construing politics instead as a domain of individual interest articulation, conflict, and adjudication rather than a domain of emerging common ground. David B. Truman, *The Governmental Process* (New York: Alfred A. Knopf, 1952; revised edition, 1971).

6. Andrew Sullivan, "Our Democracy Has Never Been So Ripe for Tyranny: The Case Against the People," *New York Magazine,* May 2–15, 2016.

7. Barber, "Democracy or Sustainability."

Chapter 6. Political Institutions Old and New

1. Jonathan Chait, "The Sunniest Climate-Change Story You've Ever Read," *New York Magazine,* September 2–29, 2015, p. 28. Chait's good news is based on new

regulations on coal, lower prices for solar, and the U.S.-China climate pact in which both sides finally agreed to do something, although what they did is not exactly "transformational" since the pact calls on China only to stop *increasing* emissions by 2030 and the United States to continue decreasing emissions more or less at the rate it has already realized.

2. The American Legislative Exchange Council (ALEC) has created a national lobby that not only pressures (and pays) state legislators to adopt their issues, but actually writes legislation that state assemblies debate and enact. See Nancy Scola, "Exposing ALEC: How Conservative-Backed Laws Are All Connected," *The Atlantic,* April 14, 2012. Jane Mayer offers a discussion of this kind of lobbying in her *Dark Money: The Hidden History of the Billionaires Behind the Rise of the Radical Right* (New York: Random House, 2016).

3. Thomas L. Friedman, "Bonfire of the Assets," *The New York Times,* August 26, 2015.

4. The extended argument is found in Benjamin R. Barber, *If Mayors Ruled the World: Dysfunctional Nations, Rising Cities* (New Haven: Yale University Press, 2013).

5. Bloomberg, "City Century."

6. Ibid, p. 116. Bloomberg argues that "the world will be shaped increasingly by metropolitan values: industriousness, creativity, entrepreneurialism, and . . . liberty and diversity."

Chapter 7. The Road to Global Governance

1. ICLEI Seoul Declaration, "Building a World of Local Action": For a Sustainable Urban Future," adopted at the ICLEI World Congress, April 9, 2015 (http://worldcongress2015.iclei.org/wp-content/uploads/2015/06/ICLEI-Seoul-Declaration_final.pdf).

2. These documents share a potent language of aspiration and hope, but represent a commitment to act rather than a catalog of agreed-upon actions, the aim of the agenda outlined here.

3. Formerly Berlin, Hong Kong, Jakarta, Johannesburg, Los Angeles, London, New York City, Sao Paulo, Seoul, and Tokyo, but succeeded in 2016 by Amman, Boston, Mexico City, Copenhagen, Rio de Janeiro, Milan, and Hong Kong, along with the continuing members Jakarta, Johannesburg, Los Angeles, London, Seoul, and Tokyo.

Chapter 8. Climate Justice

1. No one has made the linkage between the market economy and climate injustice more dramatically than Naomi Klein in *This Changes Everything: Capitalism vs. the Climate* (New York: Simon and Schuster, 2015).

2. Cited by Wen Stephenson, "On the Front Lines of Climate Justice," *The Nation,* October 26, 2015. Stephenson is the author of *What We're Fighting for Now Is Each Other.*

3. Trade-offs are inherent to the climate crisis. It is often pointed out that methane absorbs heat at a rate twenty times greater than carbon dioxide. Yet methane dissipates far more quickly than CO_2. This does not constitute a reason to be unconcerned with methane pollution, but it is another complicating factor in the final climate reckoning.

4. See Rodin, *The Resilience Dividend,* as well as the dynamic Rockefeller program run by Michael Berkowitz that Rodin initiated when she was president at the foundation. Cities do contribute to greenhouse gas emissions, but they have had little opportunity to combat the causes.

5. Shell tried to be evenhanded, inviting Rockefeller to make a strong and prudent case for resilience as something other than an excuse for avoiding decarbonization, and allowing (if that is the word) me to give a keynote speech in which I raised some of these very questions from a vantage point that I cannot call fossil-fuel friendly. For a critical study of the oil industry, see Neela Banerjee et al., "Exxon: The Road Not Taken," *Inside Climate News,* September 16, 2015 (https://insideclimatenews.org/content/Exxon-The-Road-Not-Taken); and Gary Sernovitz, *The Green and the Black: The Complete Story of the Shale Revolution, the Fight Over Fracking, and the Future of Energy* (New York: St. Martin's, 2016). Tim Flannery discusses the two books in "Fury Over Fracking," *The New York Review of Books,* April 21, 2016.

6. Nicholas Kristof points out that a hamburger requires more water for its processing and distribution than a shower. It takes 14 gallons of water to produce one mandarin orange, 12 gallons for a head of lettuce, and 3 gallons for a single walnut. An egg takes 53 gallons of water, and a pound of beef takes 1,800 gallons. Nicholas Kristof, "Our Water-Guzzling Food Factory," op-ed column, *The New York Times,* May 31, 2015.

7. As we move toward significant production of solar energy—12 percent of Hawaii's homes are solar powered, for example—progress "puts pressures on old infrastructure like circuits and power lines and cuts into electric company revenue. As a

result, many utilities are trying desperately to stem the rise of solar." Diane Cardwell, "Utilities See Solar Panels as Threat to Bottom Line," *The New York Times,* April 19, 2015.

8. See Sara Robinson, "Making Sustainability Legal: 9 Zombie Laws That Keep Cities from Going Green," *Alternet,* February 23, 2012 (www.alternet.org/story/154261).

Chapter 10. Common Principles and Urban Action

1. This would be "equivalent to eliminating around a quarter of today's carbon emissions from coal." The 2014 study was conducted by Bloomberg Philanthropies and the C40 Cities Climate Leadership Group along with the Stockholm Environment Institute, and is cited in Bloomberg, "City Century," p. 121, from which the description is quoted.

2. Ibid., p. 120.

3. C40 Blog, "Climate Action in Megacities Version 2.0" (CAM 2.0) Survey and the Carbon Cities Climate Registry, *C40 Cities,* February 5, 2014 (www.c40.org/blog_posts/CAM2). In accordance with C40 membership standards, C40 cities will continue to report to CDP. CDP has added an additional Compact of Mayors module to their questionnaire to capture all the data cities need to demonstrate compliance. CDP will then share this data with the carbon Climate Registry, the designated central repository for the Compact of Mayors, which will enable the compilation of data through existing national, regional, and global city reporting platforms.

4. The Compact of Mayors made good on trends manifesting the growing role of cities at least since 1995, when COP 3 in Kyoto established a Local Authority Subsidiary Body (LGMA); at COP 10 in Buenos Aires (2003), there was a call to recognize local governments, followed by the Montreal (COP 11) call in 2005 for the formation of the World Mayor's Council on Climate Change; at Bali (COP 13) in 2007, the Local Government Climate Session established the Covenant of Mayors, while in Cancún at COP 16 (2010), cities were formally recognized as stakeholders.

5. The data collected through the Compact of Mayors is the evidence base needed for cities to quantify the greenhouse gas impact of urban action, aiming at (1) showing national governments the extent of action cities are already undertaking, so that it might be incorporated into national level strategies or further supported through more enabling policy environments and resourcing approaches; (2) encouraging increased capital flows into cities to support local action; (3) demonstrating the

commitment of city governments to contribute positively toward more ambitious, transparent, and credible national climate targets by voluntarily agreeing to meet standards similar to those followed by national governments; (4) establishing a consistent and transparent accountability framework that can be used by national governments, private investors, or the public to ensure that we can be held responsible for our commitments.

6. "The 169 Commandments," *The Economist,* March 28, 2015; the article is subtitled "The proposed sustainable development goals would be worse than useless."

7. Pope Francis, *Laudato Si',* p. 20.

8. Rockstrom, Steffen, et al., "Planetary Boundaries."

9. The Deep Decarbonization Pathways Project (DDPP) is a collaborative global initiative to explore how individual countries can reduce greenhouse gas (GHG) emissions to levels consistent with limiting the anthropogenic increase in global mean surface temperature to less than 2 degrees Celsius (°C). Limiting warming to 2°C or less . . . will require that global net GHG emissions approach zero by the second half of the 21st century. This, in turn, will require steep reductions in energy-related CO_2 emissions through a transformation of energy systems, a transition referred to by the DDPP as "deep decarbonization." The DDPP is led by the Sustainable Development Solutions Network (SDSN) and the Institute for Sustainable Development and International Relations (IDDRI). Currently, the DDPP includes 15 research teams from countries representing more than 70% of global GHG emissions: Australia, Brazil, Canada, China, France, Germany, India, Indonesia, Japan, Mexico, Russia, South Africa, South Korea, the United Kingdom, and the United States. The initial results of this effort were published in September 2014 as "Pathways to Deep Decarbonization: 2014 Report."

Chapter 11. The Politics of Commensurability and the Challenge of Trust

1. Kaid Benfield, "A Closer Look at Siemens' Green Cities Rankings," *Grist,* July 6, 2011 (http://grist.org/cities/2011:07-05-closer-look-at-siemens-green-rankings/

2. Rémi Louf and Marc Barthelemy, "A Typology of Street Patterns," *Journal of the Royal Society Interface,* October 2014 (published online April 8, 2015; http://rsif.royalsocietypublishing.org/content/11/101/20140924).

3. Bush and Cruz quoted in Andrew Marantz, "Greener Pastures," *The New Yorker,* May 9, 2016, p. 18.

4. McKibben's constancy and moral courage were rewarded in ways deeply helpful to the struggle against climate change in 2016. McKibben's campaigning with Senator Bernie Sanders led to his appointment to the Democratic Convention Platform Committee, and the inclusion of a strong plank in the platform on climate change. McKibben was also invited to address the convention, though not in prime time.

5. Pope Francis, *Laudato Si',* p. 166.

6. Klein, *This Changes Everything,* p. 14. Klein sees market capitalism as an obstacle to climate action: "We are stuck," she writes, "because the actions that would give us the best chance of averting catastrophe—and would benefit the vast majority—are extremely threatening to an elite minority that has a stranglehold over our economy, our political process, and most of our major media outlets" (p. 26).

7. Pope Francis, *Laudato Si',* p. 129 (Pope Francis is quoting Pope Benedict's encyclical from 2009 here).

Chapter 12. City Sovereignty and the Need for Urban Networks

1. When India formally accepted the agreement during Prime Minister Narendra Modi's visit to the United States in June 2016, which meant that nations representing 55 percent of greenhouse gas emissions were in, the agreement became law. No nation can now rescind its commitments for four years.

2. Schragger, *City Power,* p. 12. Schragger believes "cities are in many ways more economically and political relevant than states or nation-states," but as a professor of law fails to recognize their full actual and potential power by treating them exclusively as economic and legal entities. Art, culture, creativity, imagination, urbanity, and community—all keys to the life of the city and its democratic character—are absent in his analysis, leading him to conclude that "why people flock to the city" remains a "mystery" (p. 42).

3. Bloomberg, "City Century," pp. 122, 124.

4. Michael R. Bloomberg and Anne Hidalgo, "Why Cities Will Be Vital Players at Paris Climate Talks," *The World Post,* TheHuffingtonPost.com, June 30, 2015 (www.huffingtonpost.com/michael-bloomberg/paris-climate-talks-bloomberg_b_7683246.html).

5. "Invest in Cities, Renewable Energy, UN Envoy Michael Bloomberg Tells Conference in India—UN and Climate Change," *UN News Center,* February 17, 2015 (www.un.org/climatechange/blog/2015/02/invest-cities-renewable-energy-un-envoy-michael-bloomberg-tells-conference-india).

6. Mayor Bill de Blasio of New York, remarks to the Vatican Climate Gathering, Reuters, "De Blasio Discusses Climate at Vatican," *The New York Times,* July 21, 2015 (www.nytimes.com/video/nyregion/100000003812515/de-blasio-discusses-climate-at-vatican.html).

7. "The Albany Pols Who Love Plastic Bags," editorial, *The New York Times,* June 6, 2016.

8. Judge Richard J. Leon ruled that "the District's understandable, but overly zealous, desire to restrict the right to carry in public a firearm for self-defense" runs afoul of the Constitution. Tamar Lewin, "Judge Blocks Curb on Gun Permits," *The New York Times,* May 18, 2016.

9. Anne-Marie Slaughter, an international law professor and former head of the Woodrow Wilson School of International Affairs at Princeton University, now the president of the New America Foundation who also has a strong record of practical public service (as policy planning director at the State Department during the Obama administration, for example), has suggested that cities can be more faithful advocates of international law than nations. Anne-Marie Slaughter, "International Law in a World of Liberal States," *European Journal of International Law* 6 (1995), pp. 503–538.

10. I argue for collective city civil disobedience against state and national laws forbidding the regulation (and even banning) of assault rifles and cop-killer ammunition in Benjamin R. Barber, "A Cities Revolution: No More Orlandos, No More Guns," *The Huffington Post,* June 2016 (www.huffingtonpost.com/benjamin-r-barber/a-cities-revolution-no-mo_b_10466758.html).

11. There is a perversion of the idea of citizen sovereignty perpetrated by the so-called Citizen Sovereignty Movement, which denies the legitimacy of federal sovereignty tout court, not because the state has failed to exercise its authority, but because it has exercised it far too efficiently in domains where Citizen Sovereignty advocates believes it lacks legitimacy: taxation, enforcing minority rights, the policing power and so forth. City sovereignty respects the legitimacy of state sovereignty but denies its efficacy. Citizen sovereignty denies the legitimacy of state power.

12. Bloomberg and Hidalgo, "Why Cities Will Be Vital Players."

Chapter 13. A Practical Climate Action Agenda

1. In 2016, following the lead of Seattle, San Francisco, Rochester, and other cities, the District of Columbia fully divested $6.4 billion from two hundred major fossil fuel companies. In the same year, the Berlin City Council voted to divest

pension funds from carbon companies. CarbonTracker.org works to align capital markets with the climate change policy agenda.

2. Mark A. Benedict and Edward T. McMahon, "Green Infrastructure: Smart Conservation for the 21st Century," *Renewable Resources Journal* 20, no. 3 (2002); www.sprawlwatch.org/greeninfrastructure.pdf, retrieved October 10, 2014. "Green infrastructure" is a generic phrase used persuasively by the Environmental Protection Agency (EPA) and other organizations. Green infrastructure can encompass many decarbonization-relevant policies, including "built infrastructure" and "waste management" (which are sometimes viewed as perspectives rival to green infrastructure, and sometimes included as subsidiary to it). The categories are relevant whether policy is directed at sustainability—preempting the consequences of climate change—or at resilience, treating with those consequences. Our use of the term here is far more specific.

3. In recent years, city growth and development in Mexico City has actually focused on cars rather than people. The Mexican Institute for Competitiveness "estimates that 42 percent of space in new developments built between 2009 and 2013 was designated for parking." New skyscrapers often reserve up to twelve floors for parking. Elisabeth Malkin, "Pollution Returns to a City Where Not Driving Is Hardly an Option," *The New York Times,* June 14, 2016.

4. Production of an automobile requires one hundred times the energy it takes to make a bicycle. A 7.2 kilogram road bicycle with a carbon frame uses 11,546,658,000 Joules of energy during its production compared to 118,284,466,000 for a 'generic car' produced in America in 2008 (Figures via WattzOn). Variations exist based on bicycle type and other external factors, but overall the bicycle has a low embodied energy. Additionally, bicycle lane construction is less energy intensive than roads for automobiles, requiring a smaller amount of space and minimal foundations. Joe Peach, "Environmental Sustainability and Bicycles: Three Reasons Two Wheels Are Great for Cities." *This Big City* blog, November 2011 (http://thisbigcity.net/environmental-sustainability-bicycles-three-reasons-two-wheels-great-cities).

5. Richard Heinberg and David Fridley made this argument in *Our Renewable Future* (Carbon Institute, 2016).

6. Penny Sackett, "Elemental Cycles in the Anthropocene: Mining Aboveground," in the Special Papers series (#520) of the Geological Society of America. This work is a considerably extended and more scholarly version of an article on the same subject that can be found in *Solutions* magazine 3, no. 3 (June 2012); www.thesolutionsjournal.com/node/1107.

7. It is estimated that by 2005, up to 33 percent of copper and 45 percent of zinc and lead were already in use aboveground. Sackett, "Elemental Cycles in the Anthropocene."

8. City Clerk of Seattle, "Zero Waste as a Carbon Neutrality Strategy," Seattle's Carbon Neutrality Initiative, September 2010 (http://clerk.ci.seattle.wa.us/~public/meetingrecords/2010/spunc20100914_7b_pm.pdf).

9. "State of the Art Waste Management in Ankara, Turkey," *Arktik,* n.d. (http://co2mpensate.com/klimaschutzprojekte/biomasse-biogas/biogas/state-art-waste-management-ankara-turkey).

10. "What Is Green Infrastructure?" *United States Environmental Protection Agency,* n.d. (http://water.epa.gov/infrastructure/greeninfrastructure/gi_what.cfm).

11. Z.-C. Zhao, "Impacts of Urbanization on Climate Change," in *10,000 Scientific Difficult Problems: Earth Science* (in Chinese) (Earth Science Committee, Science Press, 2011), pp. 843–846.

12. A value of 0 indicates that a surface absorbs all solar radiation and a value of 1 represents total reflectivity. "LEED V4 Green Associate Study Guide," *Free LEED V4 Green Associate Study Guide,* Green Building Education Services, n.d., p. 265 (http://poplarnetwork.com/news/free-leed-v4-green-associate-study-guide).

13. "Reflective roofs are more energy efficient and cooler inside and out. In Los Angeles, the average roof albedo is about 0.17. At noon on a clear summer day in the United States, a flat roof receives about 1,000 watts of sunlight per square meter. Traditional dark roofs strongly absorb sunlight and heat up the building, increasing AC energy use, cost, and associated climate warming greenhouse gas emissions." Berkeley Lab Heat Island Group, "Los Angeles Rooftop Albedo," *Berkeley Lab,* n.d. (http://albedomap.lbl.gov).

14. Julie Chao, "Global Model Confirms: Cool Roofs Can Offset Carbon Dioxide Emissions and Mitigate Global Warming," Berkeley Lab News Center, a U.S. Department of Energy National Laboratory Managed by the University of California, July 19, 2010 (http://newscenter.lbl.gov/2010/07/19/cool-roofs-offset-carbon-dioxide-emissions).

15. John Bianchi, "National Environmental Group Applauds Mayor's Plans for Extensive Pedestrian Plaza in Heart of Times Square," Environmental Defense Fund, press release, n.d.

16. Michael Grynbaum and Matt Flegenheimer, "Mayor De Blasio Raises Prospect of Removing Times Square Pedestrian Plazas," *The New York Times,* August 20, 2015 (www.nytimes.com/2015/08/21/nyregion/mayor-de-blasio-raises-prospect-of-

removing-times-square-pedestrian-plazas.html?_r=0). The eventual solution was to reserve space for the desnudas, or topless panhandlers, and other tourist attractions, preserving their First Amendment rights while protecting passersby from harassment or embarrassment.

Chapter 14. Exemplary Cities

1. Kahn quoted in Jessica Shankleman, "London Seeks to Rescue Anti-Pollution Plan from Brexit," *The Chicago Tribune,* July 9, 2016.

2. Mega-cites everywhere face the same challenge. Beijing, with more than 5 million cars for a population of 21 million people, now rations license plates through lotteries, with an annual quota far under the growing demand. Owen Guo, "Want to Drive in Beijing? Good Luck in License Plate Lottery," *The New York Times,* July 29, 2016.

3. Kamala Rao, "Seoul Tears Down an Urban Highway and the City Can Breathe Again," *Grist,* April 04, 2011 (http://grist.org/infrastructure/2011-04-04-seoul-korea-tears-down-an-urban-highway-life-goes-on).

4. For an account of the Cheonggyecheon Stream restoration, see Randy A. Simes, "The Surprising Story of Sustainability in Seoul," *UrbanCincy,* January 14, 2011 (www.urbanCincy.com).

5. "About GHP," *Green Highways Partnership,* n.d. (www.greenhighwayspartnership.org/index.php?option=com_content&view=article&id=2).

6. "East Coast Greenway: About the Greenway," *East Coast Greenway,* n.d. (www.greenway.org/about-the-greenway).

7. Zhao, "Impacts of Urbanization on Climate Change."

8. Chao, "Global Model Confirms: Cool Roofs."

9. OneNYC, "One New York: The Plan for a Strong and Just City," City of New York, April 22, 2015 (www.nyc.gov/html/onenyc/downloads/pdf/publications/OneNYC.pdf).

10. As Jarrett Murphy has written, in the long term more radical strategies of resilience like managed retreat are likely to be more successful than expensive and dubious preemptive projects like dikes and locks. "Chicago lifted itself two feet to escape the grime of lakefront life, and it reversed the course of its river so that it didn't drink its own sewage. The Greater Los Angeles Area barely had water for 100,000 people; engineering has allowed it to house more than 18 million. Much of Boston is landfill, and New York flooded whole towns to make reservoirs for its millions." Murphy, "The Flood Next Time," *The Nation,* November 2, 2015.

11. For details, see the Port of Los Angeles Sustainability Report, 2013, City of Los Angeles, Clean Agency, Inc. (https://www.portoflosangeles.org/Publications/Sustainability_Report_2013.pdf).

12. Barber, *If Mayors Ruled the World,* p. 319.

13. U.S. Environmental Protection Agency, "Region 1 EJ Showcase Community: Bridgeport, CT" (https://archive.epa.gov/compliance/environmentaljustice/grants/web/html/ej-showcase-r01.html, last updated February 21, 2016). Mayor Finch lost his 2015 reelection bid, for reasons unrelated to his environmental leadership. The impact on his pioneering environmental strategy is unlikely to be positive.

14. Rodin, *The Resilience Dividend,* p. 3.

15. Ibid., pp. 53–65. Rodin's approach has the significant advantage of being rooted in the real experience of cities and citizens gleaned from the many programs and projects of the Rockefeller Foundation. For resilience, the devil is truly in the details and the details here are more than ample. This makes for a welcome contrast with the noble but vague generalizations typical of the charters of principle surveyed above.

Chapter 15. Trust Among Cities

1. This explains how the U.S.-China bilateral climate accord asked China only to *stop increasing* emissions by 2030 while the US pledged to continue to *reduce its emissions* in the same time frame.

2. According to "LEED V4 Green Associate Study Guide," p. 12.

3. "LEED," U.S. Green Building Council, n.d. (www.usgbc.org/leed).

4. Whitney Dorn, "The Net-Zero Energy Building Challenge: Who Will Be Next?" United States Green Building Council, May 9, 2014 (www.usgbc.org/articles/net-zero-energy-building-challenge-who-will-be-next).

5. Adam Mørk, "U.N. City—A Star in Copenhagen's Harbor," U.N. City, Denmark, n.d. (http://denmark.dk/en/green-living/copenhagen/un-city-a-star-in-cph).

6. "Green Building Leading the Way in Copenhagen—Danish Architecture Centre," (www.dac.dk/en/dac-cities/sustainable-cities/all-cases/buildings/green-building-leading-the-way-in-copenhagen/>).

7. "Chrysler Building," YRG Sustainability Consultants, n.d. (www.yrgxyz.com/building-performance-operations/chrysler building-2/>).

8. "History and Development," *STAR Communities,* n.d. (www.starcommunities.org/rating-system/history).

9. Even within a single society, comparison using STAR is difficult, because these goals and objectives are defined by hundreds of evaluation measures that are

qualitative as well as quantitative, and since communities self-report. Lacey Shaver, "Alignment Between STAR and ISO 37120," *STAR Communities* blog, October 2014 (www.starcommunities.org/star-updates/alignment-between-star-and-iso-37120).

10. According to David Thorpe, these include experts in urban environmental sustainability from twenty organizations and institutions, including the African Development Bank, Cambridge University, CITYNET (Regional Network of Local Authorities for the Management of Human Settlements), the European Commission, the Ford Foundation, Harvard University, ICLEI (Local Governments for Sustainability), ISOCARP (International Society of City and Regional Planners), the Inter-American Development Bank, Karlsruhe University, the Natural Resources Defense Council, New York University, the OECD (Organisation for Economic Cooperation and Development), the Regional Planning Association, Technical University Munich, U.N.-Habitat, the University of Pennsylvania, URBACT, the Vienna Institute for Urban Sustainability and the World. David Thorpe, "What Is the Best Way to Measure the Sustainability of Cities?" Sustainable Cities Collective, April 2014 (www.sustainablecitiescollective.com/david-thorpe/243106/what-best-way-measure-sustainability-cities).

11. "The average density for [the cities studied by Siemens] is 8,100 people per square mile, which is about 2.5 times less than for Asian cities, at 21,100 people per square mile, and is also less than in Latin America (11,700) and in Europe (10,100)." Siemens AG Corporate Communications and Government Affairs, "US and Canada Green City Index," 2011 (http://www.siemens.com), 19.

12. The German index compares data from twelve major German cities. With more than 75 percent of Germany's population living in cities, this index yields more usable results than other indexes. The special German focus may also result from the fact that as a German company, Siemens offers a German-specific index.

13. Lacey Shaver concludes, "these goals and objectives are defined through hundreds of evaluation measures—both quantitative community-level outcome measures and qualitative local action measures. Communities select what they will report on." Shaver, "Alignment Between STAR and ISO 37120."

Chapter 16. Realizing the Urban Climate Agenda

1. Schragger, *City Power,* p. 259.

2. *European Charter of Local Self-Government = Charte Européenne De L'autonomie Locale* (Strasbourg: Conseil de L'Europe, Section des Publications, 1985).

3. Bloomberg, "City Century."

4. Somini Sengupta, "'Sclerotic' U.N. Security Council Is Under Fire for Failing to Maintain Peace," *The New York Times,* October 24, 2015.

5. For a discussion, see Justin T. Clark, "The City-State Returns," *Boston Globe,* August 9, 2015 (https://www.bostonglobe.com/ideas/2015/08/08/the-city-state-returns/rmuGzOlxbOzipwz8T45b1I/story.html).

6. Gerrymandering distorts democracy in the United States, concentrating heavy majorities of urban voters in single urban districts and spreading the minority vote out over a number of districts in which they have the slightest of majorities—assuring that a minority of citizens command a majority of districts.

7. For two pertinent accounts of the urban/rural dialectic, see Jim Goad, *The Redneck Manifesto: How Hillbillies, Hicks, and White Trash Became America's Scapegoats,* 1998; and J. D. Vance, *Hillbilly Elegy: A Memoir of a Family and Culture in Crisis,* 2016. I explore the collision of country and city in historical and literary depth in chapter 2 of *If Mayors Ruled the World.*

8. Following the Trump victory in the United States and the Renzi defeat in Italy, I wrote a series of essays suggesting specific measures cities led by courageous mayors might take as antidotes to populism's toxic successes, working through networks and through the Global Parliament of Mayors. These pieces include: "A Governance Alternative to Faltering Nation-States," *CityLab*, December 6, 2016; "Can Cities Counter the Power of President-Elect Donald Trump," *The Nation*, November 14, 2016; and "Cities Will Be a Powerful Antidote to Donald Trump: Social Scientist Benjamin Barber on the Emergence of a New Urban Radicalism," *Quartz*, November 15, 2016. The pieces prompted discussion in the *Washington Post*, "The West's Major Cities Are the Best Defense Against the Tide of Right-Wing Nationalism," November 22, 2016; *New York Magazine,* "Cities Will Be the Best Answer to a Trump White House," November 15, 2016.

Index